4/20/15
$24.95
BNT
AS-14
5/15

THE SECRETS OF
Q CENTRAL

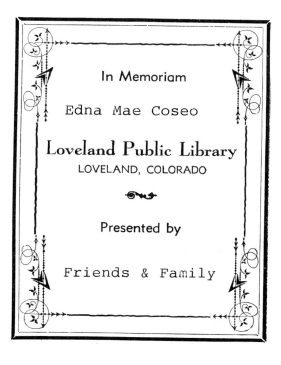

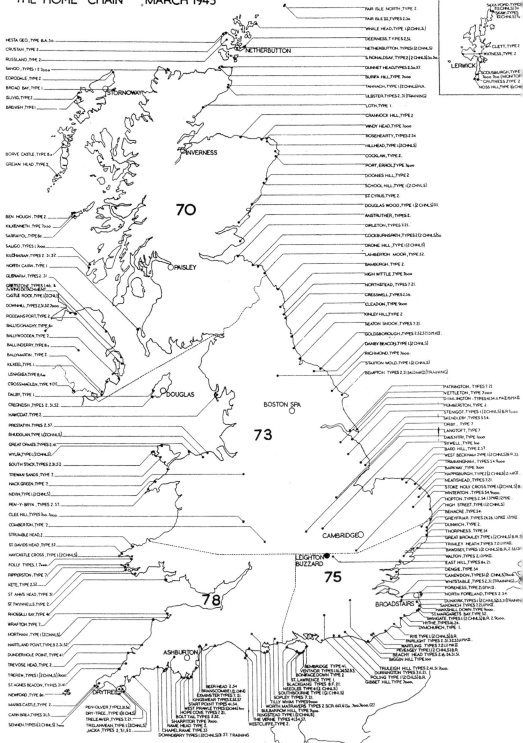

THE HOME CHAIN . MARCH 1945

All the UK radar stations ran from No. 60 Group, Leighton Buzzard, in March 1945. The circles mark Ashburton, Broadstairs, Cambridge, Boston Spa and other towns that were regional centres that would have taken over their sector if Leighton Buzzard were to be knocked out by enemy action.

THE SECRETS OF
Q CENTRAL

HOW LEIGHTON BUZZARD SHORTENED THE SECOND WORLD WAR

EDITED BY PAUL BROWN AND EDWIN HERBERT

Cover images
Front: WAAFs operating the service telephone exchange in the signals centre at Headquarters No. 26 (Signals) Group, Leighton Buzzard, Bedfordshire. (© IWM CH10243); *Back*: A commemorative piece showing Winston Churchill's well-known speech delivered at the House of Commons on 4 June 1940.

First published 2014
by Spellmount, an imprint of

The History Press
The Mill, Brimscombe Port
Stroud, Gloucestershire, GL5 2QG
www.thehistorypress.co.uk

British Library Cataloguing in Publication Data.
A catalogue record for this book is available from the British Library.

ISBN 978 0 7509 6072 4

Typesetting and origination by The History Press
Printed in Great Britain

CONTENTS

FOREWORD

The original intention of this book was to tell the local story of Leighton Buzzard and the part its people played in the Second World War, but research by members of the Leighton Buzzard and District Archaeological and Historical Society has revealed something far more dramatic, bringing together the strands of a vital secret war effort that enabled Britain to prevail – the town was the centre of a secret war machine without which the conflict would not have been won by the Allies.

One of the reasons this important history has never been recorded is that during the Cold War Leighton Buzzard continued to be a vital part of NATO's secret communications system, serving as a listening post for the radio traffic of potentially hostile forces.

What the team of researchers discovered has surprised them – even those who have lived in the town all their lives. The two editors credited on the front of the book, myself and Edwin Herbert, are part of a much larger group. Delia Gleave in particular undertook exhaustive research and wrote up the story of the evacuees and the people who served in the armed forces, including those commemorated on the town's war memorials. The team comprises Trudi Ball, Rita Barry, Paul Brown, Delia Gleave, Richard Hart, Edwin Herbert, Jill Jones and Elise Ward.

Among others whose information and memories are included are Patricia Griffin, Bill Marshall, Ludovic McCrae, Tony Pantling, Maggi Stannard, John Vickers and Raymond 'Viv' Willis. There are many sources listed in the select bibliography, but the English Heritage 2011 report on the RAF Stanbridge site and *Leighton Buzzard Observer* back issues have been particularly helpful. We also acknowledge the use of extracts for historical and educational research from the online archive of wartime memories contributed by members of the public and compiled by the BBC, which can be found at www.bbc.co.uk/ww2peopleswar.

Paul Brown
Chair, Leighton Buzzard and District Historical and Archaeological Society

GLOSSARY

AA	Anti-Aircraft
AFC	Air Force Composite (part of US eighth army command)
AFS	Auxiliary Fire Service
ARP	Air Raid Precautions (later Civil Defence) (nicknamed 'All Right Presently')
ATA	Air Transport Auxiliary
ATC	Air Training Corps
ATS	Auxiliary Territorial Service
BEF	British Expeditionary Force
CH	Chain Home – Britain's Second World War radar system
CHQ	Country Headquarters (at Woburn)
CIA	Central Intelligence Agency
EH	Department Electra House, which later became Special Operations 1, covert propaganda department of SOE in September 1941 and was given the job of preaching subversion
ENSA	Entertainments National Service Association ('Every Night Something Awful')
Gee	Navigational aid using ground transmitters and an airborne transmitter (G for Grid)
GCIU	Ground Control Interception Unit
HE	High Explosive
HQ	Headquarters
LBO	*Leighton Buzzard Observer*
LDV	Local Defence Volunteers (later the Home Guard)
Luftwaffe	German Air Force
MI5	Military Intelligence 5: The internal security service
MI6	Military Intelligence 6: The external secret intelligence service
MOI	Ministry of Information
MU	Maintenance Unit
NAAFI	Navy, Army and Air Force Institutes
NATO	North Atlantic Treaty Organization

NCC	Non-Combatant Corps (referred to as 'the Conchie Corps' by some in the Corps)
NCO	Non-Commissioned Officer
NFS	National Fire Service
Nickel	Operation involving leaflet-dropping over enemy territory
Nissen Hut	A tunnel-shaped building of corrugated iron
OTU	Operational Training Unit
PID	Political Intelligence Department
POW	Prisoner of War
PWE	Political Warfare Executive ('Pee Wee')
RA	Royal Artillery
RAAF	Royal Australian Air Force
RAC	Royal Armoured Corps
Radio 'ham'	Amateur radio enthusiast
RAFA	Royal Air Forces Association
RAF	Royal Air Force
RAFVR	Royal Air Force Volunteer Reserve
RAMC	Royal Army Medical Corps
RASC	Royal Army Service Corps
Radar	Radio Direction and Ranging
RDF	Range and Direction Finding, the early name for radar
RE	Royal Engineers
RNVR	Royal Navy Volunteer Reserve (the 'Wavy Navy')
RSS	Radio Security Service
SCU	Special Communications Unit (part of the Royal Corps of Signals)
SIGINT	Signals Intelligence Unit
SIS	Special Intelligence Service
SLG	Satellite Landing Ground
SOE	Special Operations Executive (the 'Ministry of Ungentlemanly Warfare')
TPO	Teleprinter Operator
USAAF	United States Army Air Force
VE Day	Victory in Europe Day
VJ Day	Victory over Japan Day
WAAF	Women's Auxiliary Air Force
Wehrmacht	German Army
WLA	Women's Land Army
Window	Metallic strips dropped to confuse enemy radar systems
WRNS	Women's Royal Naval Service (the 'Wrens')
WVS	Women's Voluntary Service for Air Raid Precautions (later part of Civil Defence)
YMCA	Young Men's Christian Association

PART ONE

THE SECRET WAR THAT SAVED BRITAIN

1

THE SECRET WAR

Leighton Buzzard, a small market town in Bedfordshire, was the unlikely nerve centre of Britain during the Second World War. So top secret were its operations that no outsiders knew how important it was at the time. Seventy years later, the story of its role in saving the country and shortening the war has still not been told – until now. One of its nearby satellite stations, Bletchley Park, which housed the code-breaking operations of Station X, is justly famous; but the vital part played by RAF Leighton Buzzard is still unheralded. At the time it was described in Air Ministry minutes as 'the nerve centre' of telephone and teleprinter communications.

Codenamed 'Q Central', RAF Leighton Buzzard was kept in continuous operation twenty-four hours a day throughout the war by the RAF's No. 26 (Signals) Group. According to the Air Ministry, it housed 'the largest telephone exchange in the world'. Additionally, hundreds of teleprinters were tended by staff in underground bunkers connecting the base to every theatre of war.

It is really strange that the story has never been told. Aspects of it, the evolution of the intelligence services and the activities of agents dropped behind enemy lines have been covered in various books and memoirs, but the real story of Q Central and RAF No. 60 (Signals) Group, based in Leighton Buzzard, was so top secret that all record of their vital part in the war was snuffed out by the Official Secrets Act.

A whole series of clandestine operations providing intelligence, undermining German morale and sabotaging their war machine was set up around Leighton Buzzard. Within a 10-mile (16km) radius of Q Central were Bletchley Park, the 'black propaganda' studios at Milton Bryan, the Potsgrove transmitter, the recording studio at Wavendon Tower near Simpson, and the Country Headquarters (CHQ) of the Political Warfare Executive at Woburn Abbey, Cheddington, Little Horwood and Wing airfields, the Meteorological Office run by the Ministry of Defence at Dunstable, and sites at Wingrave, Aston Abbotts and Hockliffe used by the Czech government-in-exile. Within a 25-mile (40km) radius were RAF Tempsford, the RAF Fighter Command HQ at Bentley Priory near Stanmore, the Special Communication Units for agents at Hanslope Park, and the Signals Intelligence Unit (SIGINT) at RAF Chicksands. Incidentally, the last site still has a role in psychological warfare as the HQ of the Intelligence Corps and the home of the Defence Intelligence and Security Centre.

It was from the communications powerhouse at Q Central that the British war effort was coordinated. Every vital military landline in the country was routed through the base, which was also the centre for all wireless traffic for the armed forces. Largely because it was the home of Q Central, Leighton Buzzard was also the HQ of radar command for Britain, Europe and the Far East. RAF No. 60 (Signals) Group, as it was known, was located less than 2 miles (3km) from Q Central to ensure good communications with Fighter and Bomber Command in London and hundreds of radar stations around the coasts. These two command and control centres were essential to winning the Battle of Britain and subsequently directing bombers to strike at enemy targets in Europe and further afield.

Clandestine operations directed at undermining the German war machine took many forms. Leighton Buzzard was host to hundreds of the brightest scientists, radar experts and propaganda specialists. The 'boffins' included the codebreakers

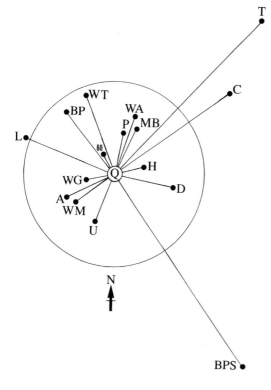

Distances and directions showing the central location of the communications centre of RAF Leighton Buzzard.

Abbreviations, clockwise from centre:

Q	Q Central
P	Potsgrove transmitter
WA	Woburn Abbey
MB	Milton Bryan studio
T	RAF Tempsford
C	Chicksands
H	Hockliffe radio station
D	Dunstable met office
BPS	Bentley Priory, Stanmore
U	USAAF Cheddington
WM	Old Manor House, Wingrave
A	Aston Abbotts Abbey
WG	RAF Wing
L	RAF Little Horwood
BP	Bletchley Park
60	No.60 Group
WT	Wavendon Tower

A WAAF wireless operator turns a message into perforated Morse tape while her companion feeds it into a high-speed transmitter at Q Central in Leighton Buzzard. (© IWM CH10281)

at Bletchley Park, some of whom were billeted in the town and cycled the 8 miles or so (13km) to work. False radio stations broadcast in German; leaflets, newspapers and forged German ration cards were dropped behind enemy lines; misleading radio signals were transmitted about non-existent troop movements; these and many more ingenious schemes were developed by units operating around the town. From the quiet countryside around Leighton Buzzard teams of specially selected and talented people, both civilian and military, established secret organisations for a new kind of warfare, called variously black, white and grey propaganda. Each part of this secret war operated on its own. Thousands of people were stationed in and around the town, all of whom had signed the Official Secrets Act.

There were those who doubted the value of some of their efforts at the time. Arthur 'Bomber' Harris, Commander-in-Chief of RAF Bomber Command 1942–45, was no fan of paper propaganda. He believed that its main effect was to equip the continent with toilet paper for five years. Dwight D. Eisenhower, a five-star general and later president of the United States, took the opposite view. He believed that it was an important contributing factor first in undermining the will of the enemy to resist and second in supporting the fighting morale of Britain's potential allies in the occupied countries.

Some efforts were heroic but others seem laughable in retrospect. For example, an attempt was made with the support of MI5 to dilute the racial purity of the SS carrier pigeon service with lower-grade British pigeons. While this plan proved flawed, pigeons were a serious part of the war effort and a lifesaving asset to airmen. Some bomber crews took them on missions in case their planes were shot down or ditched on the way home. The pigeons were then released with the coordinates of the aircrew attached in a note. Many men owed their lives to these birds.

One of the extraordinary aspects of this secret war and the range of different units in the villages surrounding Leighton Buzzard was that no one was allowed to know what the next group was doing. They had no idea what was going on behind the barbed wire and would have been arrested had they asked or, worse, answered a question about it. Seventy years have had to pass before a fuller picture of what happened in this apparent backwater of the country has been pieced together.

Airfields opened locally for other purposes were convenient for black operations and the transport of secret agents. Wing airfield, only 3 miles (4.8km) from Leighton Buzzard, was a training base for bomber pilots; but it had a satellite airfield at Little Horwood close by, which also doubled as the HQ of two Special Communications Units (SCUs). They dropped radios, spies and 'black propaganda' specialists behind enemy lines to assist resistance in occupied countries.

MI5 and MI6 relied on Q Central for communications, as did the Radio Security Service, which recruited radio 'hams' to track down German spies by pinpointing radio transmissions. Having successfully dealt with all the German spies in this way, it turned its attention to other activities, including the manufacture of covert radio sets at Little Horwood for use by Allied agents in occupied countries.

The Czechoslovak government-in-exile was at the Abbey, in the nearby village of Aston Abbotts. They ran a Czech military intelligence service from a radio station in a farm at Hockliffe, close to Q Central, to keep in touch with the resistance at home. This unit planned operations throughout the war with the British Special Operations Executive (SOE) for which the Germans exacted terrible revenge on the Czech civilian population.

When the Americans joined the war, an airfield at Cheddington was handed over to them. This too joined the secret war with activities designed to demoralise the enemy. The US Eighth Air Force dropped a billion leaflets over enemy territory, as well as secret agents to help resistance activities.

Slightly further afield but also part of the operations surrounding Q Central was the most secret airfield of the war – RAF Tempsford. Designed by an illusionist to look like an abandoned site, it was in fact the home of two special duties squadrons whose job was the delivery and retrieval of agents in enemy-occupied Europe.

It is a national story embedded in a small non-military rural market town in Bedfordshire. So important was the secret work carried out locally that the few anti-aircraft guns in the area were 'not allowed to fire at German planes unless there was a direct assault, since this would have indicated that there was something to hide in the countryside below'. Leighton Buzzard was extraordinarily lucky that the Germans never discovered what was happening because the town would have been obliterated. It was a far more important target than any industrial complex.

While it seems really strange that the story has never been told the truth is that until now it has been snuffed out by the Official Secrets Act. Looking back from this distance, it is curious that the wartime operations were so secret that the people working in them had no idea what was going on in other parts of the organisation or behind the barbed wire in the next town or village.

Meanwhile, during the conflict the town played its part in the war effort like everyone else, its inhabitants unaware of the vital secret work around them. Men and women went off to war, the town absorbed a tide of evacuees and thousands of RAF personnel, scientists, spies and propaganda experts flooded into the area. The fact that the road to Q Central was cut off for several years and the whole vast complex placed under enormous sheets of camouflage to avoid enemy attack must have seemed normal in a war of a type no one had experienced before. Local inhabitants had no inkling that RAF Leighton Buzzard also had a vast underground emergency operations room for the Air Defence of Great Britain, should the centre at Stanmore in North London be put out of action.

The following chapters describe some of the activities that took place at and around Q Central. Much of the evidence was kept in secret archives that have gradually been released over the years. Large bundles of papers were burned at the end of the war, and many of those who took part in the activities followed instructions and never talked about what they had done. But plenty of traces remain for us to gain an insight into operations without which the war could not have been won.

An important part of the story is that of the town itself. Many of the town's population of fighting age enlisted in the armed services. A large number of men from a local regiment were caught in the surrender of Singapore and suffered in Japanese Prisoner of War (POW) camps. Their story is also told in these pages.

This small country town had to play its part in other ways, for example taking thousands of evacuees from big cities including several schools. There were some tensions, but it is remarkable how they were all fitted in when all the properties of any size, and many smaller ones, were used to billet the thousands of people needed to run Q Central, Bletchley Park and other establishments. Dozens of Nissen huts were built on country house lawns to accommodate this influx.

The town was transformed by this influx of service personnel, more than half of them from the Women's Auxiliary Air Force (WAAF). Most of the people involved worked incredibly long hours and consequently needed to live it up when off-duty. On some evenings there were as many as three dances going on simultaneously. For a sleepy market town with a pre-war population of about 8,000, it was a rude awakening.

2

Q CENTRAL (RAF LEIGHTON BUZZARD) – THE NERVE CENTRE OF THE BRITISH WAR EFFORT

On 30 November and 1 December 1937, a wireless transmitter van roamed the countryside around Leighton Buzzard. Nearly two years, as it turned out, would elapse before war broke out, but the military and the government were preparing for a conflict many thought inevitable. The purpose of the exercise was to find a location for a remote reception station for radio traffic from across the world. Two large fields on the edge of Leighton Buzzard were chosen for the RAF's central communications station. The location had the added advantage of being near the existing Post Office telephone cables that ran up the spine of the country.

Landlines were considered at the time to be a very secure means of communication, much more so than radio; cyphers were used for material sent over radio, but landlines were thought to be safe enough not to warrant encyphering. Leighton Buzzard, with access to both landlines and radio, became the hub of the country's Defence Telecommunications Network, known as 'Q'. Hence RAF Leighton Buzzard became 'Q Central'.

The decision to make the town the communications centre for wartime Britain was to change the face of Leighton Buzzard, at that time a small market town with no military presence. This receiving station, originally designed for all international RAF radio signals, quite soon also housed the largest telephone exchange in the world, and became the route for army and navy radio traffic too. This turned RAF Leighton Buzzard into the secret communications hub of the British war effort. Army, navy and RAF units, as well as the secret services MI5, MI6 and MI8, all used its teleprinters and cyphers. Both Fighter Command and Bomber Command needed its signals to direct their operations.

According to Air Ministry minutes in the Public Record Office, RAF Leighton Buzzard (renamed RAF Stanbridge after the war) was vitally important to the war effort. The station contained 'the nerve centre of practically the whole of the national landline teleprinter communications and a large part of the private speech telephone system'. The minutes indicated that if the station had been put out of action the whole telephone system in the country would have been affected.

The area chosen for the new station was on the edge of town in open fields still bearing the marks of the medieval ridge and furrow crop system, next to the Marley Tile Works, which later made very uncomfortable quarters for some of the service personnel. The site had an area of 109 acres (44 hectares) and was relatively flat, rising to its highest point on the eastern boundary at 375ft (115m) above sea level. The sand and clay pits already present saved a lot of excavation in creating bunkers to protect some of the vital communications.

Initially, fearing that RAF Leighton Buzzard would not be built in time for the outbreak of war, the military took over the vast basement of the Leighton Buzzard Corn Exchange in Lake Street. Hundreds of teleprinters were installed in readiness for the coming conflict while the ballroom and other function rooms continued in use upstairs.

This arrangement continued for most of the war – dances and a cinema with a seating capacity of 450 operating on the upper floors, while the war was being run from the basement. It was not until 1943 that these original teleprinters were moved to join hundreds of new ones in the ever-expanding bunkers at the main RAF station less than a mile (1.6km) away.

The transformation from fields to a top-secret establishment entirely covered in camouflage netting took a remarkably short time. What could not be concealed in the same way were eleven radio towers 90ft (27.4m) high. Another twenty-seven towers of the same size were added later. Fortunately, although a German reconnaissance aircraft spotted and photographed these masts, the interpreters of the images dismissed RAF Leighton Buzzard as just another routine radio station. It was a costly mistake for the German war effort.

So worried were the authorities that the camouflage would not deceive the enemy that a decoy station was constructed nearby, 'a dummy block sufficiently close to the masts and sufficiently close to (RAF) Leighton Buzzard to deceive and take the attack'. A site was chosen only 500yd (460m) west of the actual station in the belief that bombers aiming at the decoy station would unlikely to be such poor shots that they would hit the real thing. Sound City Films were employed to create what was in effect a giant film set of main buildings and offices to be called 'power stations'. In addition to these were dummy huts, roads, car parks, lodges, air raid shelters and fencing. The dummy road was built by removing the top soil to a depth of 4in (10cm), the soil being made to look like darkened concrete. The 'road' was boarded with a kerb that was straight and marked with chalk. To complete the deception thirty 'old crocks' were requisitioned to put

At 11.12 a.m. on 23 September 1940 a German reconnaissance plane of
Aufklärungsgruppe 122 took a remarkably clear picture of the Q Central site, showing radio
masts, searchlight and AA positions, and the 'main building' (which was in fact a decoy
site). (© IWM HU66027)

into the dummy car park. Throughout the war a party of men maintained the decoy, repairing and repainting sections as required and living in a guard house for four constructed for the purpose.

The station itself was seen as vulnerable to air attack. Trenches were dug and Lewis guns mounted but there were no pillboxes, although at the time these were seen as the most effective form of defence. Two searchlights were set up in a field behind the Fox and Hounds (now the Flying Fox) on the A5 'Sheep Lane' crossroads (now a roundabout), as well as two 3.7in anti-aircraft guns, one on the hill at Stanbridge near the old windmill and one on Billington Hill near the Church. However, as was the case at RAF Tempsford, the guns were ordered to fire only to counter a direct assault on the station, since their use would have shown the enemy that there was something significant to protect.

The Ministry of Home Security removed all enemy aliens within a radius of 5 miles (8km) of the station and an 'unclimbable' fence was erected around the entire perimeter. No description of what made this fence unclimbable exists but it was clear that security was strict. Only those with Air Ministry passes were allowed into Q Central and the road past the RAF station, between the waterworks and the road junction near the George and Dragon Inn, was closed to the public on 30 July 1940 under the Defence Regulations.

Residents, farmers and members of the public requiring access to carry on their legitimate business had to apply for special permits from the station's Officer Commanding. The barriers were within the perimeter of the RAF site so the residential stretch at the western end remained accessible to everyone. In March 1945 the road was officially reopened, although it had been unofficially open to traffic for 'some considerable time' before that, according to the national archives.

As buildings were added to the complex the camouflage screen was continually extended. It was realised it would have been disastrous for the conduct of the war if Q Central had been disabled. To guard against this a rectangular building 'Q Central Reserve' was built in Leighton Buzzard on the south side of the High Street between the then telephone exchange and the Midland Bank (now the HSBC). It was a two-roomed, windowless concrete structure that had been refurbished and equipped with French-polished mahogany furniture. The set-up had the advantage that all the connections had their own battery power and calling was done by hand generator. Ray Parker, a 16-year-old apprentice at the Corn Exchange in 1943, had moved into Q Central in 1944, where he remained until 1954, points out that it was fortunate that Q Central Reserve never needed to be used in anger, as it was too small to have coped, with only a three-position magneto switchboard. The building was later converted by Jehovah's Witnesses into their Kingdom Hall and is still in use today.

Q Central Reserve building at the back of the HSBC bank in Leighton Buzzard High Street, now Kingdom Hall.

To keep this vital communications operation going was a major logistical headache – not least because of the need to find accommodation for all these people in a small town already packed with evacuees. There was, too, competition for space from the RAF's No. 60 Group, which also had its HQ in Leighton Buzzard, and was also growing fast. Five miles (8km) away, Bletchley Park housing Station X was also pressed for billets; and many of its personnel found homes in the town. The station expanded faster than the huts could be erected and even the WAAF Officers' Mess at Wing Lodge was temporarily appropriated as an interim measure.

The biggest problem was the airmen's home in the Marley Tile Works next to the base, described with typical British understatement as 'unsatisfactory':

Teleprinter messages came in by the mile twenty-four hours a day from all over the world and had to be turned back into plain language so they could be forwarded to the relevant department and acted on. The messages arriving at Q Central were graded for importance, 'flash' being the most urgent. (© IWM CH10298)

Ventilation and lighting are very poor, the sheds constitute a very vulnerable target and the general conditions of this sort of accommodation, although suitable as a temporary expedient, are quite unsuitable as permanent accommodation. There is no peace or privacy, since watchkeeping personnel and RAF Regiment and domestic staff are continually coming and going in this colossal barrack room. The tile-drying sheds consist of two large adjoining sheds. The former has few windows and the latter no windows but only a few lantern lights in the roof. A hot air plant installed for the purposes of drying the tiles provides a reasonably generous supply of warm ventilation, but, while this plant does maintain a tolerable standard of hygiene in the buildings, the general effect is stuffy and depressing. The sheds have a concrete floor, brick walls and open pitch roof and are structurally sound, although the circumstances in which 280 men might become casualties to a single large bomb cannot be viewed without apprehension. Running beneath the sleeping accommodation is a stagnant sewer which could be smelt within the sheds.

To solve the problem of lack of facilities, a new camp was planned to provide billets for fifty-seven WAAF officers, thirty-seven Non-Commissioned Officers (NCOs) and 956 other ranks on a field next to the tile works. There was also to be accommodation for twenty-five RAF officers, thirty junior NCOs and a further 138 airmen. The RAF Regiment, tasked with security and the defence of the base against enemy attack, had a force of eight officers, nine senior NCOs and 229 other ranks, who would also be in the new camp. Because of the lack of facilities for personnel of both sexes a large canteen was built for the Navy, Army and Air Force Institutes, better known as the NAAFI, and another just for WAAFs by the Young Women's Christian Association.

Internal War Department memos say that the idea was to build new camps to allow previously requisitioned properties to be 'released to Station X by whom they are urgently required'. The increasing importance of code breaking as the war progressed required the recruitment and housing of many more people. One of the problems the RAF encountered in building the Leighton Buzzard camp was that the site contained heaps of tiles. A check revealed approximately 351,000 tiles weighing a total of 620 tons, which had to be moved. Italian POWs were drafted in to provide the necessary labour but then sent away because of security fears.

There were many other buildings on-site. One housed the fifty-six GPO engineers essential to keep the teleprinters running. They needed a restroom, a workshop and stores. Another building used was the Leighton Buzzard Isolation Hospital. Apparently empty when war broke out, it was immediately taken over by the RAF as an 'officers' war room', although what exactly that meant is not clear. The exact location of this site is not known today and it may have been invented as a cover story.

Because of RAF Leighton Buzzard's importance a large quantity of spare equipment was stored on-site, for both its own use and the outstation's. There was also a large stock of teleprinters and stationery which required dry and temperate storage. Nearby, and sometimes in the same buildings, were explosives and small arms, which caused considerable alarm when the RAF station was inspected. None of the storage facilities reached 'regulation standard'.

In March 1944 a special inspection was carried out of 'danger' buildings and 4,300lb of explosives were found, of which a quarter were high explosives, the rest small arms ammunition and minor pyrotechnics, such as flares. Some of this was kept alongside other stores in an abandoned Nissen hut, a situation the RAF regarded as 'highly unsatisfactory'. A special building was immediately constructed surrounded by a bank 12ft (3.6m) high and of 'considerable thickness' to shield the camp from the explosives store.

Q Central: The station grew fast throughout the conflict. After fifteen months of war a report was produced describing its work. As of 1 January 1941 it was already responsible for outstations at Dagnall, Buckinghamshire, Greenford in London, the airship base at Cardington in Bedfordshire and Met Office in Dunstable. Q Central itself was divided into seven sections, each dependent on the other.

An RAF reconnaissance picture of Leighton Buzzard taken on 28 August 1940 and labelled 'secret'. The radio masts are all that can be seen of RAF Leighton Buzzard; the buildings of Q Central are all hidden under camouflage, preventing them being detected from the air.

Section One had seventeen wireless receivers of various types with six sets of equipment for transmission and reception, working twenty-four hours seven days a week. They kept in continuous contact with military commands in Aden, India, Iraq, Egypt, Malta, Singapore, Australia and New Zealand.

In addition, the wireless networks kept in touch with Ottawa in Canada when technically possible, and tracked aircraft on overseas delivery flights. There was also a link with the Home Guard in Northern Ireland, and a four-hourly schedule to be maintained with Takoradi in West Africa. Until the collapse of France, Leighton Buzzard controlled the work of two mobile wireless stations for forces there and for British forces in Norway until their evacuation. A watch was also kept on all service frequencies in use to ensure accurate tuning of all local and overseas transmitters.

Section Two was the central telephone exchange. Direct lines were provided to the Air Ministry in all its locations, the Admiralty, the War Office and to various army and navy units, all RAF Home Command and Group HQ, and a number of RAF stations. In 1941 this required eight operators on duty for efficient running, though later in the war this number increased significantly. The station was eventually connected to all theatres of war, including the Far East.

Section Three was the Central Defence Teleprinter Network Exchange, which had twenty-seven positions and needed eight to twelve operators for efficient running, depending on the time of day. There was also direct communication with all the military HQ mentioned for the telephone exchange, plus the Meteorological Office at Dunstable and Signals Intelligence Unit at Chicksands. The section also acted as a switchboard for a number of the country's secret services, including MI5 and MI6.

In November 1941 the War Office requested that Section C of MI8, the Radio Security Service based in Barnet, be connected to Q Central's switchboard. The job of the radio operators working for the Radio Security Service, including amateur radio 'hams', was to intercept radio messages from German spies operating in Britain and transmissions from their contacts on the continent. So successful were they that, as far as we know, all the German agents were either 'turned' and used to feed their Nazi controllers false information or imprisoned. The Radio Security Service was claimed by various branches of the intelligence services: it had been set up by MI5 at the beginning of the war and became Section C of MI8 before being taken over in 1941 by MI6 as its Section VIII under Brigadier (later Sir) Richard Gambier-Parry. It finally became one of the Special Communication Units (SCUs) based at Hanslope Park in north Buckinghamshire, still under Gambier-Parry, where its work was extended to manufacturing covert radio sets at Little Horwood for Allied agents in the field. This work is described later. Section

Three remained the main receiving and transmitting station for the radio traffic of MI5 and MI6, the home and foreign intelligence services throughout the war.

Section Four was the Central Teleprinter Section. The teleprinter switchboard installed at Leighton Buzzard was the first of its type. The central function of its twenty-three teleprinters was to send messages to multiple addresses. It also handled signal traffic between home and overseas commands. Nineteen machines were in use by 26 February 1941, one of which was for Station X, the radio station part of Bletchley Park. By December that year a further circuit had been installed, for the use of MI5 and MI6 tasked with intercepting German-encrypted radio traffic, which was sent to Station X for decoding.

Section Five was Transit Control, an operation designed to simultaneously regulate the radio traffic and confuse the enemy. Its Code and Cypher Section operated a 24-hour watch, while its Security Section loaded the radio channels with dummy messages to create false peaks in traffic to confuse the enemy. Transit Control ensured the whole system operated efficiently.

Sections Six and Seven were the nuts and bolts of the operation. Section Six was operated by the General Post Office and was responsible for backup generators, batteries and the maintenance of all teleprinter and telephone exchanges. Section Seven encompassed the mechanics responsible for the maintenance of all wireless and aerial equipment.

On 4 February 1940 the Central Forecast Office of the Meteorological Office was moved out of central London to the west of Dunstable (now Norman Way and Bunhill Close). It became the centre of a complex network, in partnership with the RAF 'Y' station at Chicksands Priory, that sent weather reports to the Allied military forces and to the BBC. From 1941 onwards Chicksands also served as an intercept station, picking up German traffic and sending this to Bletchley Park for decoding. The station also sent warning messages to the French Resistance immediately prior to D-Day and helped track Hitler's mighty battleships, the *Bismarck*, sunk in 1941, and the *Tirpitz*, sunk in 1944.

By July 1939 an underground standby Emergency Operations Room had also been built in case the main Fighter Command HQ at Bentley Priory near Stanmore in Middlesex was put out of action. In the event it was never required, and was used instead to train WAAFs as plotters for Bentley Priory.

The Met Office based in Dunstable, close to Q Central, was key to predicting a calm weather window to launch the 'D' Day invasion. Meteorological reconnaissance aircraft and ship's reports were received by radio and WAAF and RAF operators took them down on typewriters. The forecast was updated every thirty minutes. (© IWM CH15748)

The WAAFs attending the plotting courses were accommodated in the old Workhouse in Grovebury Road, Leighton Buzzard. Joyce Deane, who joined the WAAFs in 1940 as a Clerk Special Duties, describes the experience:

We lived in a workhouse on the first floor, access by an outside iron staircase and slept in a dormitory. The ablutions were primitive – a line of sinks with small lead bowls chained to the wall; lavatory doors didn't shut. We marched to a large building to train. The whole building was covered with camouflage netting. We learnt to plot on a replica of Fighter Command plotting table and were not allowed to stray; everywhere was secret.

Molly Ellerton, née Matthews, another trainee plotter at Leighton Buzzard, wrote the following poem, now at the Imperial War Museum:

> The thrill of going on duty in the middle of the night,
> Then months of plotting air raids in the ops room underground,
> The excitement of the news when a lost aircraft had been found;
> When the bombers crossed the ocean and the 'target was in sight!'

She describes life in the WAAFs during the war as good but admitted she had two distinct advantages: she had been to boarding school for four years just before the war and she was young and resilient.

The RAF station needed more and more people to remain fully operational. Large numbers of personnel, more than half of them WAAFs, worked exhausting shifts keeping the station running at full capacity twenty-four hours a day, 365 days a year.

At its peak, the telecommunications network comprised over 12,000 individual telegraph circuits, 562 terminal stations and 10,000 teleprinters. In early 1940 RAF Leighton Buzzard handled 3,000 phone messages daily; by 1944 this had risen to 5,000 messages, plus 30,000 teletype calls. By 1945 10,000 messages a day were being printed on 'chadless (non-perforated) tape', the volume so great that a conveyor belt was installed to transport them to a central distribution point.

The teleprinter switchboard of twenty-seven positions linked the centre with Bletchley Park, the Dunstable Meteorological Office, RAF command centres and Chicksands. The network was based mainly on Multi-Channel Voice Frequency, using amplified telephone-type circuits.

Ray Parker remembers the flight sergeant in charge of the WAAFs working on the telephone switchboard 'as a lady who wouldn't stand any nonsense from the girls but who was regarded by them as a kindly mother figure'. Doreen Luke, who worked at Stanbridge for a short period, recalled a comfortable atmosphere on watch, although she is quoted in Stephen Bunker's *Spy Capital of Europe* describing the work as 'very tiring and exacting'.

The filter room, as it was known, was used for training plotters until 1943, when No. 26 (Signals) Group took it over to house the teleprinters relocated from the Leighton Buzzard Corn Exchange. A much larger space was required for the teleprinters than even the enormous Corn Exchange could provide. This was because of the vastly increasing daily volume of traffic to direct Bomber

Command and the air cover required after D-Day when the Allied Expeditionary Air Force was protecting troops in France.

Because so much that occurred at RAF Leighton Buzzard was top secret, the purpose and contents of some of the buildings shown on contemporary plans of the site are no longer known. To complicate matters, 60 Group, based at Oxendon House, was sometimes styled RAF Oxendon to hide its true identity. After the war the new occupants of Oxendon House, a flying training group, decided they did not like this name and called themselves RAF Leighton Buzzard. As a result, a number of the 'official' photographs from Leighton Buzzard still carry misleading captions in the Imperial War Museum files. We have corrected the captions for this history.

The seriousness with which personnel took the Official Secrets Act is illustrated by a remark made by Joan Anne Williams (later Lady Llewellyn, OBE), one of the first women to volunteer for the WAAFs. After training as a codes and cypher officer, she was posted to Bentley Priory in 1940 and then a year later to Q Central. On promotion to squadron officer, she moved to the Air Ministry in London and accompanied Churchill on various trips as cypher officer, before returning to Q Central for her final wartime posting. Her obituary in the *Daily Telegraph* reports her describing to her son a nine-course 'picnic' she had attended with Churchill in Morocco. When her son said, 'I'm not the slightest bit interested in what you had to eat, I want to know what Churchill said,' she replied, 'I couldn't possibly tell the likes of you that.'

Joan Anne Williams (later Lady Llewellyn).

After the war the names of Q Central and RAF Leighton Buzzard faded quietly into history and RAF Stanbridge, as it now was, was developed as the RAF Supply Central Control System. It became the organising centre for many of NATO's supplies in Europe, housing a giant computer and controlling British military logistics worldwide. The station received the freedom of Leighton-Linslade in 1988 but in 2012 the service was outsourced to Boeing Defence UK and moved to a purpose-built centre in Milton Keynes. A single Spitfire from the Battle of Britain Memorial Flight flew over RAF Stanbridge on 21 June 2012 as the ensign was lowered for the last time.

3

NO. 60 (SIGNALS) GROUP – RADAR'S PART IN WINNING THE WAR

'I have no hesitation in saying that without No. 60 Group and Radar, the Battle of Britain, even the War itself, could not have been won,' Fighter Command Chief – Air Marshal James Milne Robb

From the beginning of the war in 1939, through the Blitz and until Germany and Japan surrendered in 1945, Leighton Buzzard was the HQ of the organisation that ran Britain's chain of radar stations. The remarkable story of what became RAF No. 60 (Signals) Group of Fighter Command is almost unknown to the people of the town and rarely mentioned even in official histories. It was so top secret that during, and even after the war, no one who had worked there was allowed to speak about what they had done.

Oxendon House, an Edwardian-style mansion in Plantation Road, was the nerve centre of radar operations for the Battle of Britain, D-Day and the campaigns in North Africa, the Middle East, Italy and Germany. No. 60 Group ran hundreds of its own radar stations and absorbed another hundred from the army and navy.

The rarely seen insignia devised for No. 60 (Signals) Group RAF – the vigilance of radar's all-seeing eye.

From small beginnings the HQ trebled in size and kept on growing. Hundreds of people were trained and dispatched all over Britain, and then the world, to run radar stations, all taking their orders from Leighton Buzzard.

Radar HQ had moved to Leighton Buzzard in a 'panic' two days before war began, because the government feared radar's then HQ on the Essex coast had been observed by German airships and would be bombed or even attacked by paratroopers. The staff originally took over a small mansion called Carlton Lodge nearby, but grew fast and requisitioned Oxendon House too. At first it was an uneasy mix of civilian scientists and RAF staff, but the work's importance meant they were soon under military discipline.

No. 60 Group was formed on 23 March 1940 to control Range and Direction Finding (later called Radio Direction and ranging or radar) stations and other radio units. By mid-September 1942 its personnel had grown to 18,000 people. There were 786 RAF officers, 315 WAAF officers, ninety American officers, about 12,000 airmen, 5,000 WAAFs and the Women's Royal Naval Service (WRNS) and 300 civilian 'boffins'. Many of these people

The HQ staff of No. 60 Group on the lawn outside Carlton Lodge, Heath Road, includes many high-ranking officers. This is a rare picture both of this unit and of the house itself, which is now gone. It is a mock Tudor house built just before the First World War and was requisitioned for radar headquarters in 1939. When it proved too small the headquarters moved to nearby Oxendon House in Plantation Road, but the senior officers continued to use it as a mess and sleeping quarters. (RAF Museum, Hendon)

began their work in Leighton Buzzard before being posted to their radar stations; others never visited the place; but to all of them Oxendon House was HQ.

Extraordinarily, this vast organisation, so vital to the war effort, has been neglected by historians. There is no doubt that without No. 60 Group directing our fighters in the Battle of Britain the survival of the country would have been unlikely. Once that crisis was over, radar was developed to become an offensive weapon, guiding bombers and invasion fleets, and became instrumental to winning the war. The importance of Oxendon House can be favourably compared with nearby Bletchley Park. The cracking of the Enigma Code at Bletchley Park is rightly credited with shortening the war, but without No. 60 Group, and the men and women who invented, improved and then ran the radar stations from Leighton Buzzard, the war might well have been lost already.

The commanders of our air forces were in no doubt about the value of No. 60 Group. Air Marshal James Robb, quoted at the beginning of this chapter, and in charge of Fighter Command, said that 5,000 enemy aircraft were shot down as a result of the information supplied by the radar stations run from Leighton Buzzard.

As for shortening the war, refinements to the radar technology allowed paths to be found through enemy defences and targets identified. Air Chief Marshal Sir Arthur Harris, Commander-in-Chief of Bomber Command, said: 'Now that something of the immense part that has been played by Radar in winning the war has been made public, it is possible to acknowledge that without the work of 60 Group Bomber Command could not have brought its task to a successful conclusion.' As Winston Churchill himself observed, 'before radar we had to rely on the same equipment to observe incoming aircraft as stone-age man – sharp eyes and keen hearing.'

So how did all this come about and what exactly happened at Leighton Buzzard? The story begins as early as 1935, when the defence establishment knew that war was coming and that Britain should prepare. In particular, the Air Defence Committee of the Air Ministry recognised the threat posed by the expanding Luftwaffe. If war were declared there was no way of stopping this huge bomber force unless some means could be found to detect them before they reached our shores. A 'boffin', the RAF term for anyone who had any scientific role, was consulted. In this case the brilliant atmospheric scientist Robert Watson Watt was able to demonstrate that radio-echoes from aircraft could be analysed to indicate their range and direction of approach.

A small research team was formed and during 1936–39 the Air Ministry Research Establishment was set up at Bawdsey in Suffolk to develop the technology eventually known as radar. Watson Watt proposed a chain of radar stations along the southern and eastern coasts to detect incoming aircraft. An embryonic chain of five stations were 'on air' in time to track Neville Chamberlain's plane when he went to negotiate with Hitler in September 1938. By April the following year there were already fifteen stations on 24-hour watch. They remained so for the next six and half years and were joined by hundreds of others. Only destruction by direct enemy air attack could put any of these stations out of commission for any length of time.

A month before war was declared the *Graf Zeppelin*, a giant German airship, was tracked flying up the coast on 2 August 1939. It was seen reconnoitring the base at Bawdsey and was later detected as far north as the Firth of Forth. It was partly as a result of this German spying mission that Leighton Buzzard's part in all this began. Only two days before war was declared on 3 September 1939 a 'panic plan' was put into operation to relocate all the research facilities away from Bawdsey in anticipation of a direct German air or paratrooper attack. The civilian 'boffins' still developing the technology were to become part of the RAF. An HQ needed to be found where the scientists could continue their work and the RAF could run the national operation.

According to papers in the RAF Museum at Hendon, the original suggestion was to send everyone to Dundee – about as far from potential German enemy action as one could get. But Edward Fennessy, the man responsible for making the decision, had different ideas. Fennessy had joined the Air Ministry Research Establishment in 1938 and assisted Watson Watt in developing the 'radio detection and ranging' system that was later called 'radar'. He also helped plan the network of coastal radar stations that would be vital to detecting and plotting the track of all incoming German bombers during the Battle of Britain.

In a brief history of No. 60 Group he wrote and deposited at the Hendon RAF Museum, Fennessy describes why he chose Leighton Buzzard for the organisation's main HQ rather than Dundee, where some of the 'boffins' had relocated. In 1939 and 1940 his primary task was to help Fighter Command defend Britain against air attack. Fighter Command HQ was at Bentley Priory, a non-flying RAF station at Stanmore in North London. Dundee was too far from the people Fennessy needed to see at Stanmore. A base in Leighton Buzzard would be near to Stanmore and have the added advantage of being close to the central communications for RAF Q Central, which was already being built at Leighton Buzzard.

The first HQ for the radar network was Carlton Lodge on Heath Road, Leighton Buzzard, opposite Sandy Lane. Known as 'Base Maintenance Headquarters', it comprised a team of sixty-seven technicians responsible for running and maintaining all the radar stations in Britain. That house, occupied on 1 September 1939, two days before war was declared, only had seventeen rooms and soon became too small. The main HQ were moved to the larger Oxendon House in nearby Plantation Road, which also proved inadequate for the increasing numbers and a score of enormous huts were built in the grounds.

Fennessy wrote about his choice of Leighton Buzzard:

It was the home of RAF Central, the hub of all Royal Air Force Communications. All telephone plotting lines from the CH (radar) stations not only went to Bentley Priory Filter Room but were duplicated to RAF Central. Here an emergency Filter Room and Operations Room would take over in the event of the destruction of Bentley Priory. So by basing Base Maintenance Headquarters close to RAF Central we had first class communications to all CH stations. We were also close by road to the vital South and South East coast (radar) stations and only some forty minutes from Fighter Command with whom we needed close contact.

Fennessy says that throughout the winter of 1939 work continued from Carlton Lodge with civilian scientists and technicians developing the radar network. Because of the system's dramatic growth and the thousands of staff now employed, a new military organisation was needed and security was vital. No. 60 Group was born. To show the importance of the unit, the chain of command was Commander-in-Chief Fighter Command, Air Chief Marshal Sir Hugh Dowding; the Director General of Signals, Air Vice-Marshal Charles William Nutting; and the on-site commander Air Commodore Arthur Leslie Gregory, a far more senior officer than would be expected at a normal RAF station, even one with several fighting squadrons.

The new unit took over Oxendon House, with civilians now in a minority and Air Commodore Gregory, supported by a large number of RAF officers, fully in command. The decision to change from a civilian to a military operation caused some conflict:

There was particularly friction between the spit-and-polish brigade and the creative boffins who wore what they liked, introduced unauthorized (but necessary) technical modifications and generally came and went as they wished. Some bitterness was engendered when boffins had to get a pass to look at their own inventions and were told they were interfering with the work. But the needs of the operational squadrons fighting the battle did act as a binding force and the RAF did not lose its top radar scientists.

Fennessy was still a civilian at that time too but transferred to the RAF, working in Leighton Buzzard throughout the war. He asked for the rank of Group Captain RAFVR and was not pleased to find after joining that he was classed as a pilot officer. His strong personality soon rectified this misjudgement, and he began his task of helping to defend Britain against the waves of bombers intended to destroy docks and industry and undermine civilian morale.

Hundreds of young women were recruited to help run the radar stations and also to act as plotters. As the radar signals came in revealing the position of enemy bomber formations and the direction they were heading from their airfields over the Channel, these plotters would show their numbers and their potential destination on a giant map.

This allowed Britain's heavily outnumbered fighter squadrons to get airborne and prepare to attack the incoming bombers, breaking up their formations before they reached their targets. In an audio interview describing these years, held in the Imperial War Museum, Fennessy says the radar could accurately tell the numbers and direction of incoming aircraft more than 100 miles (160km) away: 'Without radar development we would have been utterly incapable of fighting the Battle of Britain.'

The Germans did attack the coastal radar stations and some were damaged and put out of action. No. 60 Group's response was to send teams and equipment immediately to get them up and running again within hours. The Germans soon concluded that they were wasting their bombs trying to knock them out. Fennessy describes this as 'one of the greatest tactical errors of the war. If they had wiped out the radar stations then they would have wiped out Fighter Command as an effective force. The day that decision was made was the day Germany lost the Battle of Britain.'

One of first of the WAAFs to be recruited to No. 60 Group was Eileen Younghusband, who describes her experiences in her book *One Woman's War*. She recalls being in a group of thirty trainees aged 19 to 24, being brought to Leighton Buzzard in 1940 and put up in the former Workhouse in Grovebury Road. Next day they were taken to a camouflaged building not far from the Workhouse, where they were made to sign the Official Secrets Act and were told that they were to be trained as plotters. In the basement of another 'large building' there was a giant table the shape of southern England from North Norfolk to the Isle of Wight. This is where the WAAFs learned to interpret the radar signals and plot the path of the bombers so the RAF fighter squadrons could be scrambled to meet them.

Another plotter who left her memories is Anne Duncan. Aged 23, she had been turned down for training as a military nurse and instead volunteered for the WAAF:

It was three weeks training. We were billeted in a beautiful country house and taken by bus each day. The first day I remember because it was all covered in netting with camouflage over the top. We could not think why, but we realized then that it must be something very secret. There was a table with a map of England on it and the whole of the map was marked out in a grid. The man on the other end of the telephone, when he was reporting things coming in, he would give you, as far as I remember two letters and four numbers and from that we put down a little round disc, like a tiddly-wink. Then a few minutes later there would be another one that would be a bit further on and you could see the course gradually building up and something coming in.

You plotted everything. Everything came in over their radar tubes. We manned the things 24 hours a day. We used to be talking to these men by telephone who were sitting in the radar outstations – 'yes they are coming now, there are so many, now we can see them.'

Then one day I was on, I was connected with the station at Ventnor on the Isle of Wight, and they saw the aircraft coming for them. They were being attacked while I was actually talking to them. They were badly hit – I think several people were killed, the station was quite badly damaged.

Another recruit, Joan Fleetwood Varley, did not get to stay in a country house. Like Eileen Younghusband, she was assigned the more primitive accommodation in Grovebury Road. She said: 'The Workhouse had not been altered from Victorian days, very rough, very uncomfortable. The plumbing services were quite interesting but very poor. It was very difficult to look as you were supposed to look in uniform.'

Petrea Winterbotham describes in an audiotape held in the Imperial War Museum how she answered an advertisement in *The Times*: 'Wanted intelligent women of British origin.' After being asked at the interview if she could keep a secret, she was sent to Leighton Buzzard. 'We were told we would never be allowed to say what we did there.' Having completed her training as a plotter for No. 60 Group, she was sent to Stanmore and spent the Battle of Britain underground in bunkers moving counters across the giant map of England to show the progress of enemy bombers and fighters. It was updated continuously. Watching from a balcony above were Home Office officials. They issued air raid warnings based on the information they saw below, while RAF Fighter Command scrambled its squadrons. Sir Hugh Dowding, known as 'Stuffy' to the WAAFs, was in charge of the operation. Among the guest watchers from the balcony in 1940 were George VI and Winston Churchill.

An anonymous WAAF who spent five years at Oxendon wrote in the *Radar Bulletin*:

I remember vividly my first nervous view of Oxendon House, Plantation Road – the place that was to become the brains and the eyes of the Air Force. I remember the glorious summer weather of 1940, the smell of the pines, the ugly sturdy grey stone façade of the house that stood among them, the frightful bedroom wallpaper in our offices, lunch in the little wooden golf house across the way (bread and cheese and a glass of beer, 10d). Smallest, yet in a way the most significant of all, little painted notices dotted here and there upon the wide lawns surrounding the house, asking us to 'keep off the grass.' Keep off the grass! At Oxendon! And now every square foot of it except under the big cedar is gone – covered with huts that are still not big enough for the unending ramifications of British Radar Headquarters.

One of the most colourful characters stationed at Oxendon House throughout the war was the WAAF Squadron Officer, Baroness Elizabeth de T'Serclaes, better known as Elsie Knocker. She had been married twice, first to Leslie Duke Knocker in 1906, and then to a Belgian fighter pilot in 1916, who happened to be a Baron. In the First World War she became a legend, one of the Madonnas of Pervyse, a group of women who ran a first aid station in a cellar just 100m from the trenches. Their bravery in saving wounded men on the front line repeatedly made the headlines and they were the most photographed women of the war. Fellow volunteer May Sinclair said that Knocker 'had an irresistible inclination towards the greatest possible danger'. Among her many honours was the Military Medal for rescuing a wounded German pilot from no-man's-land.

At the age of 55 the Baroness volunteered again when the Second World War began and after a period in charge of barrage balloons was posted to Leighton Buzzard to take charge of the WAAFs working for No. 60 Group. The first record we have of her in Leighton Buzzard is her booking into the Swan Hotel on 27 March 1940. She was described as 'Baroness De Serclaes, HM Air Ministry, WAAF'. She apparently arrived at Oxendon House wearing three rows of medals gained on active service in the First World War. She was described by an administrator at Oxendon as 'a quite remarkable woman who made her presence felt throughout the radar chain, from Area Officer Commanding to the most junior WAAF'. She organised, visited and looked after the WAAFs at more than 100 radar stations around Britain. Frequently, she rode a motorbike dressed in green leathers and was billeted as befitted her rank and reputation at Ascott House, Wing, a guest of the Rothschilds.

The same anonymous WAAF above asked:

Who could ever forget that grand old lady the Baroness. She travelled fast and worked hard, a welcome figure everywhere with her rows of last war medals and her astonishingly naïve habit of telling the rudest stories in the most mixed and unsuitable company. I went on my first trip to stations with the Baroness – to Stoke Holy Cross, Dunkirk and Dover – and she showed me with sour disapproval a six-foot barbed wire fence which some misguided reformer had put up between the RAF and WAAF huts. 'Bits of blue cloth hanging on it every morning', she declared, which was a gross libel of course and she knew it. She did not believe in treating men and women like bad school children. The fence was pulled down and the invisible barrier of plain decency took its place, much more effectively at all stations.

In 1946 Baroness Elizabeth de T'Serclaes was appointed Welfare Officer of the Epsom branch of the Royal Air Forces Association (RAFA), which entailed visiting ex-RAF personnel and their dependants in homes and hospitals in and around Epsom and assisting them in any way possible: this was full-time and hard work. Fundraising was key to RAFA's work and the Baroness collected jumble, which she sold at Epsom Market. On one occasion she went to Holland to represent RAFA at a special ceremony commemorating British aircrews shot down during the Second World War, including her own son, Wing Commander Kenneth Duke Knocker, shot down over Groningen in July 1942. She was also involved in the local Lest We Forget Association, which offers entertainment and outings to disabled ex-Service personnel. She died in 1978 at the age of 93. Recently a proposal was put forward to raise a bronze statue in her honour at the place in Belgium where she and Mairi Chisholm ran their dressing station in the First World War. Contributions have already been received from the City of Dixmude, the Province of West Flanders and the Flemish government.

After the Battle of Britain, radar was developed by Fennessy and his boffins to help Bomber Command and Coastal Command, which was protecting the vital Atlantic and Russian convoys. It also provided the guidance for the US 8th and 9th Air Forces on their many missions over Germany.

With 60 Group expanding so fast there was a shortage of technicians. Many were recruited from the BBC at the beginning of the war and then experienced radio servicemen from Canada were trained and stationed in Britain. Some stayed their first night in Leighton Buzzard at the Swan Hotel in the High Street before being billeted in the town or the huts at Oxendon House.

In early 1942 a 'Gee' chain (G for Grid) of stations was built to provide accurate radar navigation as far as the Ruhr, the German industrial heartland. A southern chain, covering the English Channel and northern France, was soon followed by others. The country was divided into regions, each with its own HQ to minimise disruption if Oxendon House was attacked and destroyed. Together these chains offered a navigational service for all aircraft operating from the UK. It also provided a defensive screen from sea level to 30,000ft (9,144m).

On Sunday 4 April 1943 a large contingent of Airmen and WAAFs from RAF Leighton Buzzard and Oxendon House attended a service in All Saints' church to commemorate the twenty-fifth anniversary of the RAF. The procession was led by an RAF Band, a WAAF drum major leading, and the service was conducted by Squadron Leader Rev. Malcolm A. Clarke. On the march past in the High Street, Air Vice-Marshal R.S. Aitken took the salute.

Group Captain Raey-Jones marshalled the parade. Mr W.D. Cook (Chairman) and other members of Leighton Buzzard Urban District Council attended both the church service and the march past. Officers of the US Forces were also present. On Thursday, the actual anniversary, another large contingent of RAF personnel attended a special service in All Saints' church.

On the night of the D-Day landings, 6 June 1944, radar stations along the south coast were switched on for the first time at the direction of Leighton Buzzard HQ, just before the invasion fleet sailed. They provided navigation for the landing force through the channels swept clear of mines to the beaches of Normandy. Another group of specially built stations marked with deadly precision the ten massive German gun positions covering the approach to the beaches, enabling Bomber Command to obliterate them in the early hours of 6 June.

The RAF's No. 21 Base Defence Sector provided mobile radar units to cover the D-Day landings by locating enemy aircraft and directing patrols from the RAF's 2nd Tactical Air Force to intercept them. Four men from No. 60 Group HQ were asked to accompany one of these units (15082 Ground Controlled Interception Unit (GCIU)) to record their involvement in the American landing at Omaha Beach. Squadron Leader Norman Best, Flight Lieutenant Ned Hitchcock and two other men used a Leica to take photographs, at least two

Elsie Knocker, otherwise Baroness de T'Serclaes, the First World War heroine who was put in charge of the WAAF teams working for No. 60 Group, leads the HQ contingent on a march down Plantation Road from their HQ in Oxendon House to a service in All Saints' church, Leighton Buzzard.

WAAF drum major with a spectacular throw leads an RAF Band down North Street, Leighton Buzzard. This picture is believed to be rehearsal for the band, which led the parade on Sunday 4 April 1943 through the town to celebrate the twenty-fifth anniversary of the formation of the RAF in 1918. (© IWM CH9059)

of which have survived, despite the camera being submerged during a 600yd swim by Flight Sergeant Fulton Muir Adair of 15082 GCIU, and can be seen online under the title '60 Group Camera: The RAF at Omaha Beach' (www. therafatomahabeach.com/?page_id=1071). In the heavy fighting on Omaha Beach, 15082 GCIU suffered forty-eight casualties out of 180. Their memorial, unveiled only in 2012, can be seen at Vierville-sur-Mer. Adair was awarded the Croix de Guerre for his part in the action.

After the invasion, 'Europe 60 Group', also directed from Leighton Buzzard, landed in Normandy and established a mobile chain of radar stations. Their job was to provide navigational and bombing cover for British and American air forces operating in support of Allied armies as they advanced through France and into Germany. Radar stations were built as far afield as Iceland and North Africa.

Fennessy does not mention his own part in these events in his history of the Group. It was his blueprint for the landings that guided Allied forces through

Hundreds of personnel stationed at Oxendon House attended the RAF 25th anniversary service in April 1943. Here they march down Leighton Buzzard High Street on their way to All Saints' church. The Town Hall and Cross Keys public house are in the background.

Air Vice Marshal R.S. Aitken CBE, MC, AFC, Commander of RAF 60 Group, and Group Captain R. Reay-Jones, who marshalled the parade, at the ceremony in Leighton Buzzard High Street marking the twenty-fifth anniversary of the RAF.

the minefields in June 1944. On D–Day Plus 6 he landed in France himself, soon coming under fire from US troops unfamiliar with RAF uniforms. He was mentioned in despatches and appointed OBE in 1944. Promoted officially to group captain the following year, Fennessy took charge of all RAF offensive terrestrial radio-navigation in Britain and Europe. He died in December 2009 at the age of 97.

Although No. 60 Group HQ and as many installations as possible were placed out of harm's way, some radar stations were in the front line. The Germans took to bombing the coastal stations in 1940 and once radar became an essential part of forward operations on the continent, people were killed. The worst loss came in November 1944 when a landing craft struck a mine off Ostende in Belgium. Fourteen radar officers and 224 other personnel were lost along with fifty vehicles loaded with electronic equipment from Leighton Buzzard.

Even as late as 1944, when radar stations were tracking V1 and V2 attacks and directing bombing raids against their launch sites, the Germans retaliated with

The official caption in the Imperial War Museum describes this anniversary parade as happening 'at a town in the Home Counties'. Standing on a makeshift podium outside the bank in High Street, Leighton Buzzard, are the senior officers of the units based in the town. Air Vice Marshal Aitken takes the salute. On the extreme right of the picture is Elsie Knocker. (© IWM CH9057)

cross-Channel artillery barrages against radar installations near Dover. Several RAF and WAAF personnel were killed. A Canadian identified as 'A Dubois of Canada' described these attacks on one of the largest radar stations, Swingate, on the White Cliffs of Dover: 'I did my hands-on training at Swingate in 1942. At night, we could see the flash of the gun in France and seventy-three seconds later we would hear the crash of the shell as it landed, often with no more damage than a few dead sheep.' On one occasion the shelling was continuous for twenty-one hours.

Some No. 60 Group people from Leighton Buzzard had unexpected foreign adventures. When Russia entered the war, it was decided in 1942 to try and start a regular air service from England to Moscow. It was to be a five-year contract for a civilian service but used at first for military purposes. On the first flight, to see if such a service was feasible, Squadron Leader Charles Wilson Allsop was sent from Oxendon House to set up wireless communication between the British mission in Moscow and Leighton Buzzard. The messages between the two were to be written in a special code. Allsop was dressed in civilian clothes and had a special passport in which he was described as a 'government official' in order to protect his identity, and his mission, if he were shot down by the Germans.

He had the secret code words written on special rice paper to hand over to the British mission in Moscow. In his memoirs Allsop recalled:

> The reason for typing them on rice paper was that if we were forced down on the way, I would have to eat the rice paper, code words and all. This particular rice paper had a little sugar added to make it taste good. I sampled a bit, it was easy to eat.

The flight got through and Allsop's mission succeeded so he did not have to eat the rest of the orders.

It is clear that working for No. 60 Group, as in every other theatre of war, was stressful. Two men shot themselves during their period of service with the unit. Warrant Officer Robert Edward Lewis Boorman died on 3 December 1941 aged 47. The coroner recorded the unusual verdict of '*felo de se*' (felon of himself). He was buried in St Michael's church cemetery at Halton near Aylesbury. Flight Lieutenant James Hamilton Reid died on 14 July 1945 aged 36. After a coroner's inquest, his body was buried in Auchterarder Cemetery in Perthshire. A third

Crowds of people line both sides of the street to see the big parade. The parish church of All Saints was filled with units from both No. 60 Group and RAF Leighton Buzzard to mark the twenty-fifth anniversary of the RAF. As this unit salutes Air Vice Marshal Aitken, the back of the parade is still filing past the Cedars School in Church Square. (© IWM CH9058)

A senior officer watches the RAF returning to their HQ at Oxendon House in Plantation Road. The line disappearing into the distance shows how many RAF personnel were stationed in the town.

casualty was Leading Aircraftman Dennis William Muggridge, 21, who drowned in Spinney Pool, a Leighton Buzzard sand pit adapted for swimming, on 16 July 1944. His body was buried in Edenbridge Cemetery in Kent.

In March 1945 189 radar stations were in operation around the coast of Britain, most housing more than one type of radar equipment, and all controlled from Leighton Buzzard. With the victory of May 1945, No. 60 Group was able to end the 24-hour watch it had begun on Good Friday, Easter 1939, six months before war broke out.

In October 1945, No. 60 Group was disbanded, 'whisked out of existence with unseemly haste and little fanfare, given its indispensable contribution to victory,' according to RAF historian J.R. Robinson. Some former 60 Group personnel were amalgamated with No. 26 Group to form No. 90 (Signals) Group on 25 April 1946, under the control of British Air Forces of Occupation and Transport Command at RAF Watton.

The secrecy surrounding No. 60 Group was maintained after the war. The official history of the RAF 1939–45, by Denis Richards and Hilary St George Saunders, published by HMSO in three volumes in 1953, refers to 60 Group in a diagram of the RAF command structure, but makes no mention of its work. Bernard Howard, who was stationed at Oxendon House in 1952–53, when it was HQ No. 23 Group Flying Training Command, was completely unaware of its role in the war until he found a recent website reference, although he did wonder why the telephone exchange was so large, when he doused it with a hose during a fire drill. Coincidentally, he later worked at Plessey Radar Company when Fennessy was managing director. If only he had known at the time, he could have asked him about Oxendon House.

All that remains of this glorious episode in the town's history is a small plaque in the gateway of a house in Plantation Road erected in February 2000 to mark the 60th anniversary of the forming of 60 Group. It is near the spot where Oxendon House stood. The Edwardian-style gentleman's mansion was built in 1912 and was lived in as a private house until the RAF requisitioned it. The RAF moved out in the late 1950s and the mansion and surrounding grounds were left to decay until the site was bought in 1964 by Bedfordshire County Council for use as a remand home for boys. It burnt down in March 1974. A children's home was built on the site shortly afterwards and other parts of the grounds were sold. In 1995 the whole property was put on sale and it is now a residential home.

The last vestiges of 60 Group in 2014. Only two of the many huts remain standing, now in private ownership and used for storage. Inset: two ceramic telephone insulators can still be seen, the only evidence of the vital work carried on here.

The front cover of the last edition of the *Radar Bulletin*, the in-house magazine of No. 60 Group produced at Oxendon House, Leighton Buzzard. What the earlier magazines contained is a mystery, since the organisation was top secret. This bulletin is billed as the victory souvenir number – 1945.

The plaque erected by Leighton-Linslade Town Council in February 2000 on Plantation Road, Leighton Buzzard, to mark the '60th anniversary of the formation here of No. 60 Group, Fighter Command, Royal Air Force, the Group responsible for Britain's radar defences in World War II'. The words at the bottom are: 'In the words of the Commander-in-Chief Fighter Command, "without 60 Group and Radar the Battle of Britain, even the War itself, could not have been won."'

WHAT IS RADAR?

In the early thirties Robert Alexander Watson Watt (later Sir Robert Watson-Watt) of the National Physical Laboratory at Teddington, Middlesex, experimented with the tracking of thunderstorms using radio ranging by pulse transmission. He went on to develop the system used in the war for tracking aircraft.

Put very simply, radar can be compared to throwing a pebble into a lake: the ripples move outwards from the centre of impact and if they hit an obstacle, they bounce back. More scientifically, the primary radar principle is that electro-magnetic radiation from a known point is reflected by discontinuities in the atmosphere. The waves of radiation energy reflected are received at another known point, usually the same antenna used for transmission, detected and amplified so that an observer can tell the location of the discontinuities. The observer sees the reflected pulses as discrete blips on a radar screen.

Britain's air defence system during the war depended on a chain of radar stations that picked up signals from approaching aircraft. This information was reported to a filter room where it was plotted on a General Situations Map, corrected, collated and identified. This was then passed on in its refined form, including details of position, number of aircraft, altitude and direction, to the Royal Observer Corps, who tracked the target over land, and the various Group and Sector Operations Rooms, where it was used to direct fighter operations, give air raid warnings, assist rescue operations and instruct anti-aircraft gun crews.

Modern applications of radar include automatic doors and police detection of speeding motorists.

Oxendon House, Plantation Road, as it appeared in 1952 when still used by the RAF. On the left is one of the many huts that were placed in the grounds to accommodate the scientists, RAF and WAAF personnel needed to run No. 60 Group. Only the lodge house remains. Following the death of its owner Owen C. Wallis, in May 1939, the house had been purchased by Thomas O. Mills in 1940 but was immediately requisitioned by the RAF. (Joe Turner)

4

BLACK PROPAGANDA AT MILTON BRYAN

The role of Bletchley Park in unlocking the Enigma Code during the Second World War is well known. Apart from this vital initiative, the proximity of Q Central meant that many aspects of the intelligence war involved Leighton Buzzard and the surrounding area. Not only were these activities successful in psychological warfare, but they also played a part in the shadowy world of espionage, resistance and sabotage.

One weapon that arose from the receipt of good intelligence was propaganda to demoralise the enemy. In the early days of the war, the Woburn Abbey Riding School and Marylands, the former cottage hospital just outside Woburn, 4 miles from Q Central, housed the staff of the Planning, Editorial and Intelligence sections of Department Electra House. This code name springs from its original base, Electra House in London, site of the Department of Propaganda in Enemy Countries and the Political Intelligence Department (PID) of the Foreign Office, which provided political intelligence information in the form of a weekly summary, country by country. When the 11th Duke of Bedford died in 1940, Electra House took over the Abbey itself as its CHQ.

Life at Woburn Abbey seems to have had various bonuses. In 1939–40 there was a Lyons canteen there, supplemented by an excellent small bar, according to Robert Walmsley, one of the intelligence analysts based there. He noted that you could have five sherries or a reasonable bottle of French claret for half a crown (12½ p) – not actually as great a bargain as it first appears, since this would equate to £16 in terms of average wages at the time; gallon jars of sherry and bottles of gin enlivened the evenings. There was wonderful skating on the lake and plenty of wild life. Vernon Bartlett, a foreign affairs specialist working in Woburn House, alleged that he had been butted by a llama and bitten by a rhea in the park. Woburn, writes David Garnett in his *The Secret History of the Political Warfare Executive*, had 'more than a touch of madness'.

The future Labour Party leader Hugh Gaitskell lived for a time at 6 Leighton Street, Woburn, when he was Principal Private Secretary to Hugh Dalton, who as Minister of Economic Warfare established the Special Operations Executive (SOE) in July 1940. In August 1941 Dalton became an executive member of the Political Warfare Executive, when Winston Churchill established it as the sole authority for conducting psychological warfare. The primary directive given to the Political Warfare Executive was to 'split the German people, not to do Goebbels' job and unify them'. There was early co-operation with the Americans, and in the following month a representative of the US Foreign Intelligence Service was given a tour of the set-up at Woburn.

In March 1942 the London HQ of the Political Warfare Executive was established in Bush House, with a direct line to CHQ at Woburn. One of the 'non overt' functions of the organisation was broadcasting misleading information or 'black propaganda' to the Germans. Some of the scripts for these broadcasts were composed in the Red Room of Woburn Abbey, which today is a dining room adorned by twenty-one paintings by Canaletto.

While the BBC gained an immense reputation during the war for broadcasting 'white propaganda' that aimed to tell the truth, or at least that part of the truth the authorities felt it was appropriate to release, 'black propaganda' (nigritude) was subversive, designed to lower enemy morale by deliberate deceit. As opposed to 'grey' propaganda, where no effort was made to disguise the origin of the information, black propaganda was transmitted using radio programmes that supposedly came from German radio stations or dissident groups such as an anti-Hitler group of Waffen SS, but was in fact produced by Churchill's Political Warfare Executive.

The programmes were recorded on disc and then taken by courier to the transmitting station at Potsgrove. Recordings initially took place at Whaddon Hall in the Aylesbury Vale district of Buckinghamshire, a large manor house 500ft above sea level overlooking Whaddon Chase, from which Milton Keynes is now visible in the distance. Whaddon Hall, then isolated in the country, served as HQ of Section VIII of MI6, under the command of Brigadier Richard Gambier-Parry. In February 1940, the 'Station X' wireless interception function was transferred there from Bletchley Park. The recording process was then transferred to Wavendon Tower, a large country house on the edge of the village of Wavendon, which lies to the south-east of Milton Keynes and is best known today as the location of the Stables Theatre. It was codenamed 'Simpson's' after another nearby village south of central Milton Keynes and north of Fenny Stratford. Simpson today houses the HQ of Domino's Pizza, which for obvious reasons used to sponsor the TV series, *The Simpsons*.

In February 1942 Wavendon Tower proved too small, and a special 'black propaganda' unit was established at Milton Bryan near Woburn in a purpose-built studio complex designed by Squadron Leader Edward Halliday. It was supplemented by prefabricated huts constructed around the studio. The whole compound of about 5 acres (2 hectares) was surrounded by a 12ft (3.6m) high mesh-wire fence, patrolled by police armed with rifles and tommy-guns at night. Rooms were clad with sound-proofing tiles. The wooden poles for the forest of aerials were carted in by a Leighton Buzzard firm; on one occasion a fully-laden cart overturned, causing great consternation.

Two air raid shelters situated directly opposite the studio proved their value late one night, when a solitary German plane, prowling in the skies above Woburn Abbey, lit up the darkness with incendiaries before doubling back and dropping two high-explosive bombs. These narrowly missed two cottages at Battlesden. Had they been released a second or so later, the new Milton Bryan station would have been hit.

The head of the unit was described by the future novelist Muriel Spark as an 'immensely large and fat man who looked too big for the room', and by Labour

Studio Tower, Milton Bryan, as it looked in 2014.

politician Richard Crossman as a chief who ate, drank and looked like Henry VIII. This was Denis Sefton Delmer, a fluent German-speaker who had formerly been the European correspondent of the *Daily Express*. He was known as 'Tom' to his family and friends and referred to by those in the know as 'Seldom Defter'. The complete opposite of the normal brass hat, he was usually dressed in an open-neck shirt, an old suede jacket and, in his own words, 'shrunken and frayed grey flannels'. He was, in the opinion of the writer and brilliant forger Ellic Howe, a genius in his own field who created a new concept of psychological warfare and conducted a most impressive black propaganda offensive against the Third Reich. The object was subversion – to undermine Hitler not by overtly opposing him but by pretending to support him and thereby convey false information in apparently authentic detail.

Delmer recruited German POWs and foreign nationals of occupied countries – Hungarians, Rumanians, Bulgarians, Poles, Belgians, Dutch, Italians, Czechs, French, Norwegians, Danes, Serbs, Croats and others – mostly billeted close to Woburn around Aspley Guise. Delmer himself originally lived with his wife, Isobel, at 'Larchfield', a large red-brick house in the village known by the coded prefix 'LF'. Later they moved to 'The Rookery' at Aspley Guise, where Ian Lancaster Fleming of Naval Intelligence, later famous as the writer of the James Bond novels, was a frequent visitor (Fleming was codenamed 17F rather than 007). The house had a well-stocked wine cellar and the occupants grew their own vegetables, collected eggs from their own chickens and mushrooms from the local woods and fields, picked cherries from the garden and bought venison (like rabbit and hare, unrationed) from the Woburn estate; this all led to talk of foreigners living in luxury while Britons starved.

Initially, the low-power 7.5kW short-wave transmitting station at nearby Potsgrove was used to broadcast Delmer's black propaganda, with two transmitters named 'Poppy' and 'Pansy'. The shell of the long building housing 'Pansy' can still be seen in a field at Potsgrove, together with a smaller building that once housed the diesel generator. The staff comprised an Engineer-in-Charge, Norman Bowden, assisted by Staff Sergeant Cox and other personnel including Jack Morton, Percy Jones and Jack O'Connor. Such was the secrecy that Phil Luck, who maintained the Potsgrove transmitter in 1943, was not allowed to visit Milton Bryan and knew nothing about the operations there.

The most notable member of staff in Delmer's team was the German-born Peter Seckelmann, otherwise known as 'Paul Sanders' or '*der Chef*', a Berliner with a long aristocratic nose who had a talent for mimicking the upper-class drawl of the Prussian *Junker* officer class. He came to Britain in the late thirties and enlisted in the Auxiliary Pioneer Corps, becoming a corporal and thus gaining his other nickname of 'the Corporal'. American jazz, news and even pornography were used to attract attention to the broadcasts and the language was 'rude and robustly offensive'.

Transmitter 'Pansy' and generator building at Potsgrove in 2014.

During his first broadcast ever on 23 May 1941, inaugurating the *Gustav Siegfried Eins* radio station, *der Chef* referred to Churchill as the 'flat-footed bastard of a drunken old Jew' in order to win credibility with German listeners. A later programme recounted in lurid detail a supposed orgy involving a German admiral, his mistress, four sailors and an army helmet used as a chamber pot. Sir Stafford Cripps happened to hear one of these broadcasts and described it as 'the worst foul and filthy pornography'. Delmer defended the crudity as necessary for attracting German listeners and making them susceptible to the hidden propaganda. *Der Chef* continued to broadcast until November 1943, when the last tranmission gave the impression that a 'Gestapo unit' had stormed the station while it was broadcasting and *der Chef* had died in a hail of bullets.

One of the radio announcers at the Milton Bryan studios was Agnes Bernelle. As 'Vicki', a 'lanky, spotty-faced, just-out-of-school girl of 15', she became the German Vera Lynn and *Atlantiksender* station's 'Sailors' Sweetheart'. Her signature tune, 'A Smooth One', became familiar to thousands of German listeners. Using intelligence gathered from German newspapers and letters to and from U-boat POWs in Britain and Canada, she was able to send out birthday greetings not only to the U-boat crews themselves but also to their families, thus showing familiarity with her 'dear boys in blue'. Born in Berlin, Agnes had escaped to England in 1938 with her playwright and theatre-owner father, Rudolf Bernauer. Many of her family who remained in Germany died in the gas chambers at Auschwitz. When her father was asked by the Secret Service if he knew of any girls who could broadcast in German, he suggested Agnes. Ironically, despite her wartime success, when she later applied for a job with the BBC German Service she was told her voice was totally unsuitable for German audiences. Instead she went on to become a leading West End actress.

Another member of staff at Milton Bryan was the 20-year-old 'Peter W'. He was paid £350 a year for his broadcasting work, 32s 6d (£1.62) deducted each week for board and lodging. He describes in his memoirs, held at the Imperial War Museum, how he worked in a small office equipped with a radio receiver, desk, typewriter, telephone for internal communication and a padlocked box with a slit on top for waste paper. He became one of the voices of *Wehrmachtsender Nord*, pre-recorded like other early ventures. This programme was intended to spread rumours among the German armed forces of desertion, malingering, currency hoarding, suicides, sabotaging the rationing system, etc. He relates that some German POWs who worked at Milton Bryan pretended they were still in POW camps and received promotions and decorations while actually working for Sefton Delmer.

One black station, *Blauwvoet*, sought to demoralise Belgian collaborators, and another directed propaganda specifically at Rommel's *Afrika Korps* in North Africa. Fake broadcasts were also made in Italian. After Otto Skorzeny of the German special forces rescued Mussolini on 12 September 1943, and Mussolini spoke on the radio about the incident, he was followed on-air by Captain Ottolenghi of Milton Bryan, who parodied his talk with the words: 'First speaka da Mussolini, then speaka da I'.

As at other wartime establishments, the staff organised occasional entertainments. The guard-in-charge at Milton Bryan was Sergeant Copperwheat. He doubled up as *compère* at concerts and one evening announced that 'Miss Woodley will now oblige with "One Fine Day" by Madame Butterfly.'

The food at Milton Bryan was notoriously bad: Sefton Delmer once exclaimed that 'When it's toad in the hole even Father won't say Grace,' a reference to Father Eisenberger, a noted gourmand not over-concerned with the quality of the food. With some irony Delmer described the canteen as 'a tower of Babel, as dark-haired beauties from the Balkans flirted with my fair-haired German prisoners over toad in the hole, powdered egg omelettes, spam fritters, soya bean sausages and other irresistible delicacies'. It must have been a welcome relief when for three months a new canteen manager showed what could be done by skilful and imaginative use of wartime rations.

In late 1942 a more powerful 600kW medium-wave transmitter named 'Aspidistra' (after Gracie Field's song 'The Biggest Aspidistra in the World'), located at Crowborough in Sussex became available, which made it possible to mimic German domestic stations. From the studios at Milton Bryan, programmes were transmitted by a dedicated landline to Crowborough and broadcast to all parts of Germany. After von Stauffenburg's attempt to blow up Hitler in July 1944, news of which had been intercepted at Milton Bryan, the intelligence team began to implicate as many German officers as they could over the airwaves, to sow panic and confusion among the enemy. Another use of the new transmitter was to record the instructions of German ground control one evening and then re-transmit them the following evening, so that German nightfighters were sent to the wrong destinations.

In November 1942 the production of black 'paper propaganda' was consolidated in a unit under Ellic Howe (also known as Armin Hull), credited by Delmer as a counterfeiting genius. The unit could supply anything from a few letterheads and rubber stamps with authentic German 'Gothic' lettering to several million German ration cards. Paper propaganda took many forms. In 1944, for example, an office was set up at Milton Bryan to produce a newspaper, *Nachrichten für die Truppe* ('News for the Troops'), for release to the retreating German forces in Europe. The newsletter included troop movements – invariably towards Berlin – and clever 'official' denials of profiteering by Nazi leaders. The layout was undertaken at Marylands and the final copy printed at the Manchester Street offices of the *Luton News*, 5 miles away. Packed into special containers, up to 2 million copies of each edition were then dropped over the German forces by the USAF. The newspaper was printed from 25 April 1944 to Issue 381 on 7 May 1945.

One further aspect of the paper war of words was the dropping of 20,000 satirical leaflets over the Luton and Dunstable area during 1941. This was to test the power of black propaganda on one's own people. These leaflets portrayed Hitler pouring scorn on local people's fundraising for new warships. How these leaflets were dropped, perhaps supposedly from German planes, is not clear. However, they were an outstanding success and achieved the purpose intended by the black propagandists. Over £1,400,000 was contributed – enough, in fact, for three destroyers – but there was disappointment that none of the ships was named after a local town.

The black propaganda machine also had particular success demoralising U-boat crews. American jazz was not permitted to be broadcast in Germany, but the young U-boat crews and indeed thousands of young soldiers in the Wehrmacht greatly enjoyed it. Many popular German recordings reached Milton Bryan from neutral countries such as Sweden; and programmes containing jazz with a German flavour were transmitted in order to attract German listeners to tune in to the disinformation channels. Delmer even managed to use one of the top German bands, under Henry Zeisel – captured *in toto* in North Africa by the Eighth Army – to record American hits with German lyrics. So, even as POWs, the German musicians were still playing for the Wehrmacht. Carefully screened German refugees and even U-boat POWs with inside knowledge were used to add authenticity to the announcements.

One such former U-boat man was *Flotillen Oberfunkmeister* (Flotilla Chief Radio Petty Officer) Eddy Mander from Hamburg: his technical knowledge, contacts and ease with racy lower-class slang made him invaluable. He was originally a member of the Nazi party, but as a POW he was identified as full of bitterness towards his own officers and the Nazi leaders, and he was recruited to Delmer's unit. It was easy to suggest to U-boat men over the air that the young officers, with their hunger for decorations, were the natural enemies of their crews; and Eddy Mander knew how to make sure that the grousing would go straight to the heart of the older petty officers.

On one occasion, he sent out a series of cypher signals, over a British naval transmitter, which helped the Royal Navy to lure two U-boats into a trap. He was also adept at thinking up new grouses that might be included in the black propaganda broadcasts. And when it came to suggesting how U-boat men might delay the departure of their ship – and thus prolong their lives – by petty and unattributable acts of sabotage, not even the Royal Navy experts were more imaginative. Mander's career in black propaganda lasted rather less than a year: he was returned to a POW camp after accusations from fellow POWs at Milton Bryan about homosexual advances. After the war, he returned to Germany, where he soon died – supposedly from tuberculosis but more likely at the hands of vengeful former U-boat men.

Another character, who arrived at Milton Bryan in January 1942, was the actress Margit Maass. Her particular role was to play a spiritualist medium who had received messages from deceased members of the German armed services, using names and addresses from the intelligence services, for transmission to their bereaved families. Margit's problem, however, was that she found it difficult to read the scripts without bursting out laughing and thus spoiling the recordings. She fared better as 'Red Joanna' on the 'German Workers' Station', which was cited by a British report of 26 November 1942 as responsible for passive resistance by workers in Italy and also for two cases of successful resistance by the crews of German ships.

A deserter from the Waffen SS, *Obersturmfuhrer* Zech-Nenntwich, known locally as Herr Nansen, was a 'bright-eyed, bouncy, rosy-cheeked young cavalry officer' recruited by MI5. He was installed by Delmer in the 'nouveau Tudor mansion' Paris House in Woburn Park, where he ran a short-wave transmitter pretending to be an anti-Hitler group operating behind German lines. As Delmer did not completely trust him, he was not allowed to be part of the Milton Bryan operations. On his return to Germany, he tried to redeem himself by calling the staff at Milton Bryan 'worthless traitors and collaborators'.

Examples of the black propaganda newspaper *Nachrichten für die Truppe*, edited at Marylands by Harold Keeble, who later became features editor of the *Daily Mirror*. News stories (some bogus, some more or less accurate), sports results and even pin-ups were used to create a paper that it was hoped would demoralise front-line German troops by countering the heavily censored official news they received from on high.

The black propaganda broadcasts from the renowned *Soldatensender West* (Soldiers' Radio West), which began on 24 October 1943 as *Soldatensender Calais* with the aim of demoralising the Wehrmacht troops in France prior to invasion, ended, according to Delmer in his autobiography *Black Boomerang,* at 5.59 a.m. on 14 April 1945. The occasion was celebrated by a fancy-dress party in the printing shop at Marylands, when for the very first time security restrictions were relaxed to allow the Marylands and Milton Bryan people to mix and visit each other's offices. Ian Fleming joined the gathering and John Gibbs, whose presses at the *Luton News* had rolled off 160 million copies of *Nachrichten*, arrived in a suit made up of front pages from the newspaper printed on calico.

Just after VE Day on 8 May, Delmer called the staff of Milton Bryan together and said: 'Tonight just pull the plug, don't say goodbye.' All the records from the studio were burnt in 40-gallon drums. It took several days. Phil Luck, quoted in the book *Dunstable and District at War*, records seeing a single log book from Wavendon Tower that had escaped the destruction. At about the same time, the production of the bogus German newspaper *Nachrichten für die Truppe* stopped. Programmes for the last two days of broadcasting, actually 29–30 April rather than the date quoted by Delmer from memory, were recorded on large discs by an engineer at Milton Bryan, Harold Robin, who kept them in his attic for thirty-seven years; they are now held at the Imperial War Museum.

Despite expressing the view that 'propaganda is something one keeps quiet about', Delmer's autobiography, published in 1962, contains significant details about his work at Milton Bryan. His decision to break his self-imposed silence was, he said, due to his belated realisation that he had, by his black propaganda work and subsequent silence, contributed to the development of the myth that

Denis Sefton Delmer, recruited by PWE in 1940 from Lord Beaverbrook's *Daily Express* to run the Milton Bryan black propaganda unit, was in 1944 given the title of Director of Special Operations against the Enemy and Satellites.

the 'good Generals' and the 'good *Wehrmacht*' were always against Hitler. He didn't want Germans masquerading as 'anti-Nazis' to be restored to power in the new European community.

His work gained further publicity on Monday 8 January 1962, when he was the subject of a 'This is Your Life' TV programme, introduced by Eamonn Andrews. Agnes Bernelle, the U-boat 'Sailors' Sweetheart', was one of the guests.

The studio, huts and the guard house at the entrance are still there today and the tower, now surrounded by a protective fence, was used by the Scout Association until it became unsafe. It is estimated that preservation work would cost up to £100,000. It is surprising that despite its importance Delmer's work at Milton Bryan has not received a more permanent memorial.

In the final months of the war, most of the Political Warfare Executive staff were transferred to the Psychological Warfare Division of the Supreme Headquarters Allied Expeditionary Force (SHAEF) under General Eisenhower. A small unit was retained within the Political Warfare Executive until 1946 with the job of 're-educating' German POWs. Dangerous, ardent Nazi POWs were classed as 'black', anti-Nazis classed as 'white' and regular non-political prisoners as 'grey'.

Commemorative plaque at Milton Bryan unveiled by Lord Howland (later the 15th Duke of Bedford) on 4 September 2002. Funding for the plaque, under an initiative set up by John Taylor, was provided by the Bedfordshire Tourist Board and Emerson Valley Combined School in Milton Keynes.

5

BEHIND ENEMY LINES – THE ROLE OF THE RAF AIRFIELDS AT LITTLE HORWOOD, WING, WOBURN AND TEMPSFORD

RAF Little Horwood

Little Horwood, 2.5 miles (4km) south-west of Winslow in Buckinghamshire, was a satellite of RAF Wing, a large bomber-training base 3 miles (4.8km) from Leighton Buzzard. Like other locations in the area it too became part of the secret war. Earlier work carried out at Q Central and at Barnet trapping German spies in Britain was adapted to help Allied agents in France.

Rapid expansion was required in this key area. The main aim was to support the French Resistance and promote sabotage behind enemy lines in preparation for the eventual D-Day invasion. Much of the equipment made at Little Horwood was for the agents dropped behind enemy lines from the base at Tempsford, 25 miles (40km) away on the other side of Bedfordshire. Little Horwood also housed 92 Group Communications Flight, whose job it was to jam enemy signals.

The Royal Signals presence at Little Horwood included two Special Communications Units (SCUs) undertaking the design and manufacture of covert radio equipment and the training of operators. Nissen huts provided accommodation and the centre was known as 'Gees' after the London building contractor, George Gee, who had built Whaddon Hall, 3.5 miles (5.6km) north-east of the airfield, as a hunting lodge and stables in the 1930s.

The airfield became operational in September 1942 and eventually had three concrete runways connected by a perimeter road from which tracks led to thirty heavy bomber dispersal sites. It was a joint-use field shared by the RAF and the army. Many new buildings were needed to accommodate the large signals team manufacturing wireless sets and people to train those who would operate them in the field.

An RAF map shows a communal site on the east side of Spring Lane, a WAAF site on what is now Willow Road and Nook Park, and living quarters on four other sites – one south of Little Horwood Road (now The Close), and others in fields off Singleborough Lane, Pilch Lane and Winslow Road. Nissen huts survive on the west side of Winslow Road.

In early 1944 another SCU arrived in connection with D-Day plans. As the war progressed, the three SCUs worked hand-in-hand with two other 'cloak and dagger' government departments, the SOE and PID, to drop spies and specialist 'black propaganda' operatives behind enemy lines. These groups were also largely responsible for aiding and abetting the work of the French Resistance, a large part of which was the development of espionage equipment, which had become a key element of the co-operation between the special communication units and special operations.

A motorised transport section at Little Horwood provided round-the-clock transport between the various communications outstations, comfort ranging from the luxury of a Humber Snipe estate to the wooden slatted seats of an Austin 'Tilley' Utility. A small fleet of American Packards provided transport for high-ranking officers' use and for training personnel on mobile exercises involving the use of radio.

Due to its location near Whaddon Hall, SCU No. 1 made use of a large perimeter building at Little Horwood. It was a well-kept secret, even from most of the staff on the airfield, that this building was the main workshop for the manufacture and assembly of a critical tool in the war, the clandestine radio known as the 'Paraset', so-called because it could be dropped by parachute with or to SOE agents operating in the field. The radio was only about 4 × 6 × 9in (11 × 14 × 23cm) and weighed 5lb (2.3kg). In its metal box it looked like a tin of biscuits. However, with accessories, battery and mains adapter, weighing a further 12lb (5.6kg), the assembly needed a small suitcase or panniers on a bicycle to carry it.

Development of this Morse radio had started in London with two prototypes, which then moved to Whaddon Hall for small-scale trials and manufacture by the SCUs. After tests proved very successful, the radio was put into production under strict secrecy. The Paraset was one of the first successful miniaturised radio sets used for espionage and other activities behind German lines during the war. It operated on two emission frequency bands (3.3–4.5Mhz and 4.5–7.6Mhz) selected by a toggle switch. Parasets were supplied to resistance groups in France, Belgium and Holland. However, in the later stages of the war, the German direction-finding units became adept at locating Parasets – due to a design fault, the reaction-type receiver sent a detectable signal back up the aerial. By 1944 the average duration of a Paraset transmission had to be kept to less than three minutes per frequency to avoid detection; after D-Day use was mainly restricted to Norway.

Training sites for clandestine radio operators included Howbury House at Waterend in Bedfordshire, where spies learned to use portable ground-based beacons, and two Buckinghamshire houses, Grendon Hill near Aylesbury, where they practised encoding messages for transmission, and Poundon House, where radio emission and reception were taught.

By December 1944, station personnel at Little Horwood included 126 officers, of whom two were WAAFs, and 1,402 other ranks, including 296 WAAFs. Despite its importance, the airfield lacked the comforts of Wing: it had no camp cinema, no library, no tennis or squash courts and no NAAFI; its redeeming feature, according to Neville Hockaday, one of the aircrew, was a general lack of 'bull'. Its assigned function, and a useful cover for the clandestine activities, was basic training in flying and gunnery. Wing, a larger training operation, concentrated on advanced flying training and certain elements of ground training.

In wartime any airfield could be pressed into service for any purpose and the airfield saw the spectacular arrival of American aircraft on 23 January 1943, when a flight of badly shot up B-17s from the 91st Bombardment Group made a

The clandestine Whaddon Mark 7 (paraset) radio developed by the Secret Communications Unit. These devices were made at Whaddon Hall from 1941 and at Little Horwood from early 1943 onwards.

forced landing. The casualties from the raid on L'Orient in Brittany were rushed urgently to the Royal Buckinghamshire Hospital at Aylesbury.

Flying ceased in November 1945 and the airfield was closed early in 1946. The site of Little Horwood airfield is at present owned by Greenway Farms, who are redeveloping part of the site as a business park. The remains of the airfield buildings are split into two sections, to the south is the current 'Roddimore Stud' area and a larger complex to the north side called 'Greenway Farm' (which was the location of 'Roddimore Farm' in the war).

RAF Wing

The parent base for Little Horwood was the RAF airfield at Wing in Cublington Parish in Buckinghamshire, 4 miles (6.4km) west of Leighton Buzzard. The main role of RAF Wing was to train bomber crews. However, several other types of operational flying were undertaken from the airfield, ranging from dropping leaflets over towns near the French coast (called Operation Nickel) to diversionary raids or feints whereby aircraft flew along the French coast to keep the enemy radar busy in a certain area by dropping metallised strips ('Windows') while the main force flew inland to bomb their targets. Later, the airfield also housed No. 60 Group's Radar Navigation Test Flight, two Wellingtons that were testing the Gee chain, and after the end of the war it became a reception centre for POWs returning from the Far East.

Following their basic RAF training, those accepted for aircrew first went to an Elementary Flying Training School, where they received their first taste of flying and its hazards. Often initial training was on Airspeed AS.10 Oxfords or De Havilland Tiger Moths. After a period of ten weeks or so, the would-be pilots usually had a spell at a Service Flying Training School and then graduated to an Operational Training Unit (OTU). The one at Wing was No. 26 OTU, where the pilots undertook further training. In the case of Wing, the aircraft were mostly Ansons or Vickers Wellington bombers.

The land at Wing was chosen for the airfield as it was an area of relatively flat ground with no particular threats from hills or other obstructions. All Saints' parish church in Leighton Buzzard with its tall spire and the mound called Castle Hill in the centre of Wing formed useful landmarks. The site was between the roads that ran from Wing to Cublington and Wing to Stewkley on church land forming part of Glebe Farm.

In order to avoid groups of farm buildings, the airfield was built in an unusual shape, which meant that one of the three concrete runways was very short at 1,160yd. Construction was completed by 18 March 1942, with the main runway running approximately west-south-west to east-north-east towards the

Wing–Stewkley road. The first aircraft to land was a Tiger Moth piloted by the newly appointed base commander, Group Captain S.M. Park. It also had five hangars and various other buildings. It was a large base, the station personnel comprising, by December 1944, 168 officers, including ten WAAFs, and 2,319 other ranks, of which 501 were WAAFs.

The Piper Tomahawks of USAAF No. 1684 Bomber Defence Training Flight moved to Wing from Little Horwood in July 1943, the unit's six fighter aircraft being used to give fighter affiliation training to bomber crews. Hawker Hurricanes replaced the Tomahawks in March 1944 and the flight remained at Wing until it was disbanded on 1 August 1944.

Various buildings were constructed at Wing, such as offices, a canteen, rest rooms, blast shelters, radio and telegraph rooms, training blocks, a church, a gym, a squash court, a rugby and football field, a cinema, tailors, barbers, shoemakers, post office and stores. Some remain today. The cinema showed two films every night except Thursdays, when those stationed there were encouraged to showcase their talents. There were thirteen sites for personnel to live in, each with up to twenty Nissen huts, toilets and one or two air raid shelters. The WAAF had their own site closer to Wing village. There was also a hospital close to Cublington. The airmen cycled to local pubs in Wing, Stewkley, Cheddington and Cublington in the evenings or local railway stations on their days off, so they could go home and see their families. WAAF driver Joan Greenacre (née Webdale) recalls regular runs from Wing to the Post Office at Leighton Buzzard or the railway station. 'On the return journey from the main Post Office it was commonplace to stop at Faulkners in Linslade for their delicious bread rolls whilst the girls on the coal lorries would be expected to organise a sack of coal for their hut.'

Dances and other entertainments were often put on at the airfield and weekly hops in Wing Village Hall brought many couples together. Besides the camp cinema there were Entertainments National Service Association (ENSA) shows once a fortnight, the station's own band, a music circle and a Brains Trust, as well as visits to The Bell in Leighton Buzzard and regular dances at the town's dance halls, the Old Vic in Hartwell Grove, the Corn Exchange and in the ballroom above the Co-op. After such events, a forbidding WAAF sergeant would stand outside the WAAF quarters at Wing. 'We were lucky indeed to get a goodbye kiss,' said disgruntled Ron Stotter.

The station cricket team had full use of the excellent cricket ground belonging to Ascott House, and one player met his future wife there when she was acting as scorer. Much of the fruit and vegetables supplied to the station came from Ascott House gardens. One of the gardeners, Wally Willis, called the cabbage field 'Hitler's Patch', which it remains to this day.

Neville John N. 'Hock' Hockaday served at Wing as a staff instructor. In his own words, penned at Brighton in 1980, he was one of the people described

The remains of the WAAF quarters at Wing today. To the left of the tower were the ablutions and baths of the sergeants and airwomen.

by the *Daily Herald* as the 'Little Men' of Britain, those who left their humdrum civilian jobs in 1939 and later fought first the Germans and the Italians, and then the Japanese, until the war was won; who then returned to their jobs and proceeded to enjoy the freedom for which they had fought, for themselves, for their children, and for their grandchildren. One of his favourite memories of Wing was being woken up each morning by a charming WAAF bringing him a cup of tea.

From time to time, the RAF members at Wing, both instructors and trainees, clearly found it necessary to let off steam. Hockaday describes how he and a group of officers from Wing were invited to a social evening at Oxendon House in Leighton Buzzard, the HQ of No. 60 Group. One of the party, a New Zealander named Leonard 'Lenny' Chambers, became bored and decided to liven up the place by setting fire to the anteroom curtains. Fortunately for Britain's radar's defence, it was extinguished before it got out of control. The president of the Wing Mess Committee had a few words to say to Chambers on his return. He promptly alleviated any further boredom by joining an operational unit, which turned out to be 617 Squadron, the 'Dambusters'. He flew on the famous raid in 1943 with Flight Lieutenant H.B. 'Mickey' Martin in 'P for Popsie' and was awarded the DFC.

On 30 May 1942, a 1,000-bomber raid was carried out on Cologne, followed by later raids on Düsseldorf, Bremen and Essen. Despite being a training unit, aircraft from Wing were used in these raids and five bombers were lost, depriving the Training Unit of several experienced instructors. A decision was therefore taken that training units would not henceforth be used in future large-scale raids, except for the occasional 'maximum-effort raid' or leaflet dropping or decoy raids that did not enter enemy territory.

The Training Unit was disbanded in March 1946. Maintenance Command assumed control, and Wing airfield was eventually closed to flying on 4 April 1956. In 1960 the site passed out of the authorities' hands as an operational airfield and into history. The remains of the buildings and runways can be seen from the public footpaths that cross the area. A comprehensive account of wartime operations at RAF Wing is given in *Wings over Wing* by Michael Warth, published in 2001, which includes many photographs of aircrew trained at the base and a list of the accidents, sadly extensive, that occurred during training at the airfield.

RAF Woburn

The landing strip at Woburn, the 'Country Headquarters' of the Political Warfare Executive, ran uphill from west of the Abbey to the lake at the southern end of the park. At 1,400yd (1,280m), it was one of the longest grass strips in the country. It was used from July 1941 onwards to store operational aircraft, since the RAF's maintenance units (MUs) were becoming overcrowded due to the increase in aircraft production. Supermarine Spitfires, a few Hawker Hurricanes, Vickers Wellingtons, Handley Page Halifaxes, Miles Masters and Magisters, de Havilland Tiger Moths, and Fairey Battles, Swordfish and Albacores and later the odd Avro Lancaster were accommodated there; surplus Short Stirlings were also housed there until the latter part of 1947, when they were scrapped on the site of the modern animal park.

From 1943 the airstrip acted as a Satellite Landing Ground for the Maintenance Unit based at RAF Little Rissington. As well as providing maintenance facilities, it fitted new guns, gun sights and wireless equipment. The aircraft were heavily camouflaged in storage and the site was guarded by local army units or the Home Guard. Facilities and buildings were deliberately kept to the minimum and the site did not appear on any aeronautical chart, since it would have made a plum target for the Luftwaffe.

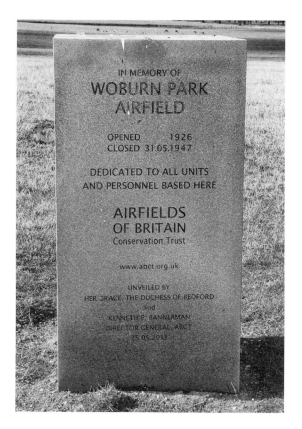

The granite Airfield Memorial at Woburn Abbey, unveiled on 25 May 2013. The dedication was organised by the Airfields of Britain Conservation Trust.

Lettice Curtis in *The ATA Angle* recalls the use of the airstrip in preparation for D-Day:

> Later in the war the Stirling played a major role with the airborne forces. No 34 Satellite Landing Ground (SLG) in Woburn Park was a storage unit for the maintenance base at Little Rissington. Here aircraft were prepared for towing gliders for use in the invasion of France. Flights in and out of satellites were normally undertaken by maintenance unit pilots but in the run-up to D-Day, because of pressure of work, No 1 Ferry Pool of ATA (Air Transport Auxiliary) at White Waltham was called on to help. The strip at Woburn to the west of the Abbey ran slightly uphill. There were at that time a number of trees alongside the strip under which aircraft were parked to make them less obvious from the air. During 1944 at Woburn, Stirlings were fitted with yokes and release gear for towing gliders. Before D-Day some 200 Stirlings were stored there under trees, 25 on half-an-hour alert and another 50 on one-hour alert for delivery to operational squadrons. After D-Day many returned. Some were broken up on site, others were flown into No 12 SLG in the park of Beechwood House, an

aircraft graveyard near Harpenden where the most notable feature was steeply rising ground which only became apparent on the last stage of the approach, calling for much last-minute pulling back of the control column.

After the war, Kenneth Halton, billeted in Leighton Buzzard, passed by Woburn Abbey:

> There the great open expanse of parkland was filled with lines of Lancaster and Stirling bombers. RAF mechanics were swarming all over them and stripping out the machine guns, instruments and engines. Our mighty air force which the enemy had failed to destroy and now we were destroying it ourselves.

RAF Tempsford

Further from Leighton Buzzard than some of the other secret establishments, Tempsford was used by agents trained at the various facilities located nearby, such as the Czechs at Hockliffe. The local villagers were quite ignorant of its wartime role until many of its secrets were released from the Official Secrets List in 1998. The German intelligence services suspected its existence but never pinpointed its exact location.

Some of the pilots who flew SOE agents from RAF Tempsford. The barn where agents were kitted out before their missions is visible in the background. (Collection of Norman Didwell)

Another group of Tempsford pilots. Flying Officer Leslie Frank Stannard of 138 Squadron is third from the right. He had been headmaster of Beaudesert Boys School, Leighton Buzzard, from 1937, until he joined the RAF, and Principal of Leighton Evening Institute.

The airfield was 2.5 miles (3.7km) north-east of Sandy, Bedfordshire, near the village of Everton. It was perhaps the most secret airfield in the war. Designed by the illusionist Jasper Maskelyne, the airfield was constructed in 1941 to give the impression of an abandoned site. It was home to the two Special Duties Squadrons: No. 138, which dropped SOE agents ('Joes') and their supplies into occupied Europe, and No. 161, which specialised in landing and pick-up operations.

One Linslade resident, Bill Marshall, a teenager at the time, remembers that Westland Lysander aircraft could sometimes be seen flying low over the fields near his home:

We noticed these aircraft could land and take off in the small fields around Linslade and were puzzled at what they were doing. Many years later I learned of the Special Operations Executive (SOE) operating out of Tempsford in north Bedfordshire from where they dropped and collected special agents in and from France. The planes we saw were obviously pilots practising their skills, particularly personnel delivery and retrieval by landing in occupied Europe.

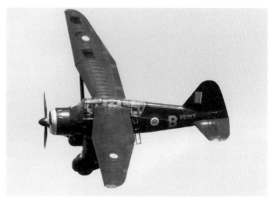

The two-seater Westland Lysander as used by No. 138 (Special Duties) Squadron RAF, based at Tempsford, for dropping agents and supplies over enemy-occupied territory. (Nigel Ish)

Reunion at Gibraltar Barn, Tempsford. In the centre is Barbara, partner of Wing Commander Forest 'Tommy' Yeo-Thomas, GC. According to Sophie Jackson's *Churchill's White Rabbit*, published by the History Press in 2012, Yeo-Thomas was the inspiration for Ian Fleming's James Bond. (Collection of Norman Didwell)

Engraved on one of the memorial plaques in Church Square, Leighton Buzzard, is the name of Flying Officer Leslie Frank Stannard of 138 Squadron. He had been headmaster of Beaudesert Boys School in the town before enlisting in the RAF. His last mission was as navigator in a Handley Page Halifax flown by Flying Officer Harcourt H. McMullan RAAF. From Tempsford, Stannard set the course for Sønderholm Heath about 4 miles (6km) east of Nibe, where twelve containers and two packages were dropped for the Danish resistance. On the return journey, the Halifax was intercepted and shot down 37 miles (60km) north of Thisted by Feldwebel Klaus Möller of *12/Nachtjagdschwader 3*. All onboard the Halifax perished. Stannard's body was washed ashore near Tjärnö and buried in the local cemetery.

In his book *Two Eggs on my Plate*, Oluf Reed Olsen recalls that the thing all the people remember about Tempsford, apart from the lively entertainments put on by the young ladies of the base, was the signal of two fresh poached eggs, at that time almost unobtainable, carefully laid on a plate before a particular person at supper time. That meant he or she was on the programme for that night's operations.

In all, 995 SOE agents and resistance fighters were dropped into enemy-occupied territory and a similar number were brought out, at a cost of 126 aircraft that failed to return. Gibraltar Barn, where SOE agents were kitted out prior to departure, has been renovated and still stands as a memorial to these extremely courageous men and women.

In January 2013 the Tempsford Parish Council adopted a proposal put forward by one of the villagers, Tazi Husain, to set up a monument in memory of the many female secret agents who had flown out of Tempsford and other RAF airfields to aid resistance movements in Europe. The Tempsford Memorial Trust was established with Baroness Crawley as patron to carry out the proposal, and on 3 December 2013, HRH Prince Charles unveiled the monument, a white marble column set on a black granite plinth, sited at the corner of the Millennium Sanctuary opposite St Peter's Church Cemetery. Engraved on the sides are the names of seventy-five female agents, fifteen of whom never came back. The face of the column has a blue glass mosaic disc depicting a white dove flying beneath a full moon. The sides of the plinth also commemorate the men and women of the two special duties squadrons based at Tempsford and those of the Allied secret services who gave their lives in the clandestine struggle for freedom.

6

PIGEON POST – ALIVE AND DEAD

One of the stranger attempts at subterfuge during the war concerned the humble homing pigeon *Columba livia domestica*. Both sides regarded homing pigeons as a vital part of the war effort because they could carry messages quickly and effectively when enemy lines could not be crossed and radio signals would be intercepted. Pigeons were trained to fly from a remote location to the home loft containing food and water. Initially, the sites were only a few hundred yards apart, but as the birds got more experienced the distance was increased. This helped the pigeons to find the home loft, even from long distances, apparently by using a mixture of magnetic, olfactory and visual signals.

At the outbreak of war there were 57,000 pigeon fanciers in Germany; the Nazis exerted central control by placing the German Pigeon Federation under the SS. Each pigeon fancier was required to have a certificate of political reliability, which excluded anybody with Jewish connections. When France was occupied, every German Intelligence section there was required to have a collection of homing pigeons to send messages. Heinrich Himmler expressed a personal interest in the eugenics of pigeon breeding and ordered any sub-standard lofts to be wiped out.

In 1944 Flight Lieutenant Richard Melville Walker, who headed a special section of MI5 and operated on the border between ingenuity and insanity, became convinced that 'Nazi' pigeons were infiltrating Britain by parachute, high-speed motor launches and U-boat. Some of the experts he consulted even claimed to be able to identify pigeons with a German accent. After unsuccessful experiments with three Peregrine falcons to bring down such intruders, which unfortunately killed twenty-three British pigeons, Walker devised a scheme whereby second-rate British pigeons were disguised as German pigeons with new leg rings, and released over France. Being in poor health, it was hoped they would prefer to go to the nearest French loft rather than attempt the long journey back to Britain. Sooner or later, Walker reasoned, they would be found out and if Himmler could be persuaded that mixed breeding was going on, he might be led to apply his final solution to all the pigeons. MI5 were enthusiastic and soon Walker had 350 double-agent pigeons at his disposal. The delivery method used was to throw eight birds in a hessian bag from a plane, a cord releasing the neck of the bag. However, interrogation of the German officer in charge of pigeons after the war established that the British pigeons released under Walker's scheme were never detected: they generally blended in with the local population and, like so many ex-combatants, made new lives for themselves, their wartime exploits unsuspected and unsung.

Pigeons were also used by SOE agents flying to France from Tempsford airfield, near Sandy. Initially they were carried in socks with a hole cut in them so the pigeon could poke its head out. Later they were housed in brown cardboard boxes with a hole in the top. The pigeons were to be released to confirm agents' safe landing. Pigeons carrying a small spying kit attached to their leg were also dropped over villages using handkerchief-sized parachutes, the idea being that the French resistance could send them back with information about the occupying forces in their area. A red chequered cock bird named 'Commando' was awarded the Dickin medal, the animal VC, in 1942 for making three trips with an SOE agent to France. In August and September 1943, 138 Squadron, flying from Tempsford, despatched 510 pigeons and 161 Squadron dropped 550. In the same period in 1944 the respective figures were 922 and 592.

It is estimated that altogether over 16,000 pigeons were released over occupied Europe during the war. Only some 1,800 (11 per cent) arrived back. One pigeon was returned with the message: 'I had the sister of this one for supper. Delicious. Please send us some more.' Others were returned with obscene messages attached; yet more were undoubtedly shot on sight by the German coastal defences. However, those that got through carried valuable information. Some even brought out microfilmed maps or copies of documents.

Pigeons saved lives in one particular way. If a bird was taken on board a bomber or reconnaissance aircraft and the plane had to ditch in the sea, the pigeon could then be fitted with a note giving the coordinates and released to fly back to its home base. A blue-chequered hen bird called Winkie won the Dickin for saving the crew of a Beaufort Bomber in this way. When the plane went down in the North Sea 100 miles (160km) off the coast after a raid on Norway, it flew 129 miles (200km) in atrocious weather conditions, covered in oil, fuel and salt water, to its base at RAF Leuchars. Other recipients of the Dickin won in this way were the Irish 'Paddy' and the American 'GI Joe'.

The USAAF, having already toyed with a plan to attack Japan using Mexican bats carrying incendiaries, also realised the value of pigeons. Pat Carty records in *Secret Squadrons of the Eighth* that the 406th Squadron of the 492nd Bomb Group was given the task of dropping carrier pigeons over Germany. Some were dead and had false messages attached to their legs, the aim being to deceive anyone who picked them up. The live ones, assuming they survived the descent as well as the flight, were intended to allow Resistance groups to send messages back. The squadron's first 'Pigeon Raid' was on 11 April 1945, when three aircraft, in addition to their normal bomb-load of leaflets, carried ten live pigeons each. During the rest of the month 340 pigeons were dropped over suitable targets in groups of ten.

Among the more bizarre secret work carried out at RAF Leighton Buzzard was the experiment testing the effect of keeping pigeons underground in battle conditions. Despite scepticism from the RAF Pigeon Service, permission was given on 11 May 1943 for a brick shelter with a corrugated-iron roof to be built for this purpose, but to save costs the shelter was neither underground nor covered with earth. No report has been traced of the results.

At the end of the war the Air Ministry Pigeon Section, headed by the enthusiastic Wing Commander William Dex Lea Rayner, devised schemes involving pigeons carrying explosives or spreading bacterial infections. These were not taken up and by 1948 the Air Ministry had concluded that there was no further justification for the military use of pigeons.

Our feathered friends are remembered today at the 'Animals in War' Memorial in Park Lane, London, and at the 'Pigeons in War' display at Bletchley Park, consisting of memorabilia collected by the late Flight Lieutenant Dan Humphries.

Columba livia domestica, the humble homing pigeon.

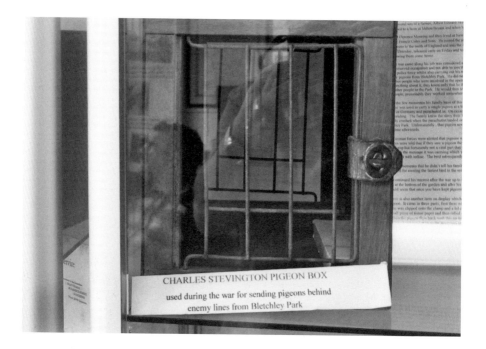

The Charles Stevington Pigeon Box used during the war for sending pigeons from Bletchley Park behind enemy lines. (Reproduced by permission of the Bletchley Park Trust)

Pigeon container and parachute used to drop pigeons over enemy territory.

7

THE CZECH RESISTANCE AND THEIR AGENTS

Although officially the Czech government-in-exile was based in London, the president of Czechoslovakia, Dr Edvard Beneš (1884–1948), and his household took over The Abbey, Aston Abbotts, in Buckinghamshire from 13 November 1940, where he converted the billiards room into an office. His Chancellery was housed at The Old Manor House, in nearby Wingrave. The location proved useful because nearby Addington House became the command post for his military intelligence and the Czech resistance.

The Czech government-in-exile was intially able to keep in radio contact with resistance groups in its occupied homeland via a military radio station located at Woldingham, near Redhill in Surrey. After September 1942 it was moved to College Farm (previously known as Rectory Farm and now called Trinity Hall Farm) a mile (1.6km) south-east of Hockliffe near Leighton Buzzard and Q Central. The station at College Farm was built by the SOE; radio equipment was supplied by the SCU at Whaddon Hall.

One of the clandestine missions planned by the Czech military intelligence service and the SOE was 'Operation Anthropoid': the assassination on 27 May 1942 of Reinhard Heydrich – the *Reichsprotektor* of Bohemia and Moravia – by two agents, Jan Kubiš and Jozef Gabčík, flown out from Tangmere in December 1941. Although the Czech resistance expressed severe doubts about the value of the mission, President Beneš insisted it went ahead. This action indeed proved counter-productive in that the savage reprisals, accounting for at least 5,000 lives and the complete destruction of the Czech villages of Lidice and Ležáky in June 1942, discouraged further active resistance. Operations mounted by 138 and 161 Squadrons based at RAF Tempsford led to heavy loss of life among Czech agents, and from 1943 onwards SOE attention was concentrated on Western Europe.

The station at Hockliffe had a permanent complement of ten radio operators, four mechanics, a cook and a foreman, under the command of Captain Zdeněk Gold, supervised by Colonel František Moravec, who was based at Addington

Wartime photo of Aston Abbotts Abbey, occupied by Edvard Beneš, the exiled president of Czechoslovakia, 1940–45. It was leased for 20 guineas a week from the owners, Captain Harold and Beatrice (née Shaw) Morton, who moved into the chauffeur's cottage for this period. Major Morton, as he became as Commander of 'C' Company of the 1st Buckinghamshire Battalion of the Home Guard, was invested after the war as a commander in the Czech Order of the White Lion (third class) by the president, in recognition of his services.

The Abbey today. View from the public footpath in the field by the Abbey, 2012.

Present-day photo of the Old Manor House at Wingrave, built by Hannah de Rothschild between 1876 and 1878, now called Mount Tabor House and Stables, having been converted into apartments in 2008 (this photo was taken in 2012). It was leased from Mary Eveline Stewart-Freeman, Countess of Essex, for £20 a week from November 1940. The Czechs kept rabbits and three pigs here, the pigs named after the French collaborators, Pétain, Darlan and Daladier.

House. One of the personnel, Sergeant Antonin Šimandl, designed both the Marjánka receiver and a transmitter named after him for use by agents in the field. Most of the men were recruited from the Signals Unit of the Free Czechoslovak Army. The station's buildings consisted of three Nissen huts painted black, with inadequate hot water and poor sewage disposal. At first there was a complete lack of any fencing as protection from cattle in the adjacent fields; a full 2 miles (3.2km) of barbed wire were put up later. Off duty, Hockliffe personnel used every opportunity to get away from this spartan existence by taking trips to the cinema in Dunstable or travelling to London via Luton.

The station used twelve aerials each for transmitting and receiving. The transmissions kept the Czech military intelligence service in contact with the Czech resistance, wireless operators on SOE parachute missions, and embassies in unoccupied countries like Sweden, Portugal, Turkey and Switzerland. The site was just off the main cable route from Fenny Stratford to London and information was sent by post office teleprinters at Q Central to SOE HQ in Baker Street, the Broadway HQ of the Special Intelligence Service (SIS), and the Czech offices in London.

Generally speaking, relations between the Czechs and the British authorities were good. The Czech soldiers providing the president's bodyguard at Aston Abbotts had their own football team called 'The Carpathians' and played matches against local teams. However, one British official expressed concern about the lack of security covering the radio traffic being transmitted from Hockliffe, particularly to German-controlled areas, and he suggested an unannounced check of the station and its logbooks. From April 1944 security was tightened and the Czechs were required to provide English-language copies of all signals despatched.

The station was fully operational until June 1945, when most of the men were flown back to Czechoslovakia in Douglas Dakotas, landing at Pilsen, which had been liberated by the US 3rd Army under General George Smith Patton in May that year. One man, Frank Kousal, then ran the station alone for communications between London and Prague until March 1946. Nothing remains of the site today.

In September 1943 President Beneš paid £150 for the building of a brick bus shelter for the local communities of Wingrave and Aston Abbotts. This was formally opened on 15 April 1944 and is located at the cross roads on the A418, approximately halfway between the two villages. During the president's stay, there had been no bus service to connect the two villages, and he had often seen people standing in the cold and wet weather at the open bus stop. The shelter was built by Messrs Fleet & Roberts of Wingrave and still stands today. The plaque below was first proposed in 1948, but was not put up until 1992.

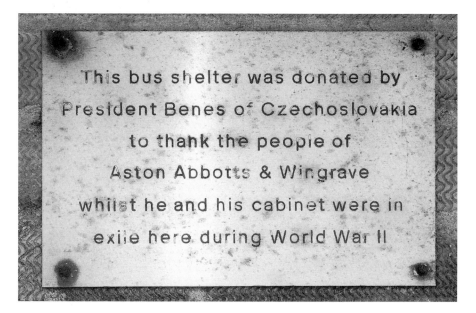

Plaque at the bus shelter on Leighton Buzzard to Aylesbury Road (A418).

Dr Beneš also donated £100 to Aston Abbotts Primary School on Cublington Road, which was used after the war to buy a climbing frame for the grounds. When the school closed in 1972, it was moved to the village recreation ground. On 31 October 1943 a lime tree was planted in the drive of The Abbey at Aston Abbotts, to commemorate the twenty-fifth anniversary of Czech independence, regardless of the German occupation; the tree can still be seen today. The last of Dr Beneš's staff left The Abbey on 17 March 1945. Fifty years later, on 8 May 1995, a commemorative plaque was placed in Wingrave church.

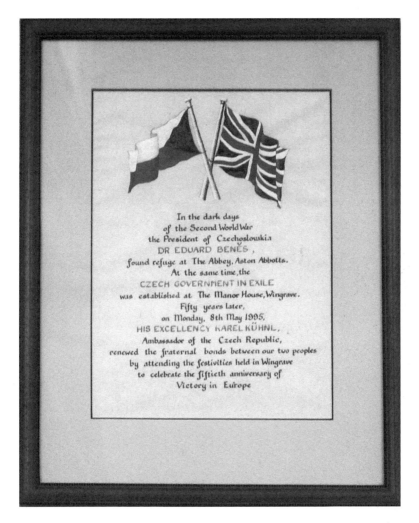

In the dark days
of the Second World War
the President of Czechoslovakia
DR EDUARD BENEŠ ,
found refuge at The Abbey, Aston Abbotts.
At the same time, the
CZECH GOVERNMENT IN EXILE
was established at The Manor House, Wingrave.
Fifty years later,
on Monday, 8th May 1995,
HIS EXCELLENCY KAREL KÜHNL,
Ambassador of the Czech Republic,
renewed the fraternal bonds between our two peoples
by attending the festivities held in Wingrave
to celebrate the fiftieth anniversary of
Victory in Europe

The plaque in the parish church of St Peter and St Paul in Wingrave, put up on 8 May 1995, the fiftieth anniversary of VE Day, to commemorate the Czechoslovakian president's stay at The Abbey, Aston Abbotts, during the war.

8

THE UNITED STATES JOINS THE 'PSY' WAR

After 1942 many American servicemen were stationed at RAF Wing's satellite station at Cheddington. According to local reports they became a familiar sight in Leighton Buzzard, shopping, drinking and socialising. The operational missions flown by these American crews were some of the best kept secrets of the war. Like many of the activities around Leighton Buzzard, they contributed significantly to the success of the intelligence and propaganda activities constituting what was called 'Psy War'. There was also a major USAAF effort to supply the resistance movements in Europe under the codename 'Carpetbagger', a derogatory term used in the American Civil War to denote Northerners carrying travel carpet bags who came in during the reconstruction period after 1865 with the intention of exploiting the South.

Cheddington airfield was set in a triangle of countryside bounded by the picturesque villages of Cheddington, Long Marston and Marsworth. Arthur Reeve, the tenant farmer of Church Farm in Marsworth, had all his land requisitioned in 1940 to enable the airfield to be built. The first he knew of this was when he received a letter from the Air Ministry instructing him not to plough his land. William Boyer Southernwood, the owner of Great Farm in Long Marston, received a similar letter.

Clearly something important was up, as at that time it was a chargeable offence not to cultivate land normally used for that purpose. Within days workmen arrived to remove trees and level the ground and in June 1941 the George Wimpey firm was employed to finish the earth-moving and construct the necessary runways and buildings. In *Voices from an English Village*, Ivan James recalls working on the airfield with a horse and four-wheeled trolley. After the runway had been concreted, it was protected with muslin cloth while it dried. James would drive the horse and trolley along the runway with a spraying machine hooked onto the trolley that released a dark liquid to camouflage the shining white surface, which was otherwise highly visible from the air. He remembers that two members of the Indian armed forces, who were deaf and dumb, worked the pump and the spray. He explained what needed to be done by sign language and says he got on well with them. Several other Indians worked on the site and many more Irish, billeted in rows of huts on Cheddington Lane, Long Marston.

Eventually the airfield consisted of three concrete and woodchip runways, two of them 1,400yd (1,280m) by 50yd (46m), with four hangars and thirty-seven dispersal bays. By 1944 the personnel comprised 558 officers and 1,993 enlisted men. The airfield was assigned the call sign 'Hillside Two' because it was overlooked by Southend Hill, some 150ft (46m) above the elevation of the airfield.

The first unit to move in, on 15 March 1942, was not American, but 121 officers and other ranks of No. 26 OTU RAF. Initially, Avro Ansons and Vickers Wellingtons were flown from Cheddington, but after several accidents the RAF decided the airstrip was not suitable for training purposes and the unit left on 3 September 1942. Cheddington was handed over to the USAAF and the first American unit to land there, after crossing the Atlantic, was the 66th Squadron of the 44th Bomb Group ('The Flying Eightballs') with their B-24D Liberators.

The crew of the strangely named aircraft 'This is it men/Bama Bound-Lovely Libba', part of the United States 36th Bomb Squadron operating out of Cheddington from June 1944. This crew was involved in jamming enemy radar, carrying out 'spoof' raids and disrupting enemy tank communications, as well as other Black Ops designed to confuse the enemy.

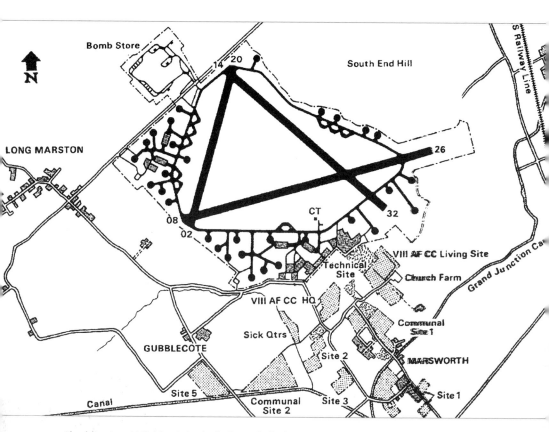

Cheddington Airfield as it looked after refurbishment in 1943.

However, despite being newly constructed, the airfield proved unsuitable for immediate operational use and remedial work had to be undertaken. On 31 December 1942 the Americans left and the RAF trainers returned with Vickers Wellingtons, together with Hotspur gliders and Master II tugs of 2 Glider Training Squadron. The station was then closed for major improvements. On 16 August 1943 Cheddington was re-opened and converted to USAAF station 113, a base for the Consolidated B-24 Liberator bombers of the Combat Crew Replacement Center. This helped to release frontline aircrew to fly combat missions. In June 1944 specialist USAAF units began to perform special operations from Cheddington, such as leaflet drops and electronic countermeasure missions. The 422nd US Night Leaflet Squadron moved to Cheddington in June 1944, becoming the 858th Bomb Squadron of AFC and in August 1944, the 406th. It flew B-17 Flying Fortresses, later replaced by B-24 Liberators.

From 27 June 1944 to 13 March 1945, the squadron carried out 378 sorties and dropped over 1.1 billion 'black' and 'white' leaflets. On D-Day leaflets were dropped on French villages warning the inhabitants that bombers would blast districts that had access roads leading to bridgeheads. It was noteworthy and gratifying that the majority of Germans surrendering after the landings carried safe conduct passes dropped by the B-24 Liberators.

The earliest method of leaflet distribution consisted of throwing broken bundles from plane windows, doors or bomb bays at high altitudes. In the first leaflet raids, pilots of B-17s and B-24s threw out leaflets when they were several miles away from the target area, trusting that the wind would do the rest. Some of the paper propaganda dropped over France was picked up in Italy! A slight improvement came when leaflet bundles were placed in cardboard boxes and released through a trapdoor attached to a bomb shackle, but it was not until the enterprising Captain James L. Monroe, technical liaison officer of the 422nd Bombardment Squadron, invented the leaflet bomb that a fully satisfactory method of distribution was

Loading a Monroe leaflet bomb. The aircraft is a B-17G Flying Fortress with its remotely operated 'chin turret' for forward defence visible on the left.

found. The new bomb, which came into regular use on the night of 18/19 April 1944, was a cylinder of laminated wax paper 60in (1.5m) in length and 18in (46cm) in diameter. A fuse that functioned at altitudes of 1,000–2,000ft (300–600m) ignited a primer cord which destroyed the container and released the leaflets. Each bomb could hold about 80,000 leaflets, which would be scattered over an area of about one square mile. The Special Leaflet Squadron's bombers were modified to carry twelve leaflet bombs, two more than a regular bomber.

The planes also dropped secret agents and psychological warfare items such as safe conducts and forged ration cards that disrupted local economies when lucky bearers flooded stores asking for scarce food supplies. As indicated in the section on Milton Bryan, a 'German' daily newspaper, *Nachrichten für die Truppe*, a joint Milton Bryan-Marylands production printed at Home Counties Newspapers in Luton and Waterlow's in Dunstable, was also dropped. At one stage as many as 2 million copies were printed a night, the proofs being sent to Sefton Delmer at Milton Bryan every morning for approval.

In June 1944 the 36th Bomb Squadron (formerly the 803rd), arrived at Cheddington, flying specially equipped B-17 Flying Fortresses and B-24 Liberators that were able to jam early warning radars and telecommunications and deceive the enemy into thinking that other (non-existent) bomber formations were assembling. This early form of electronic warfare was very effective in disrupting German activities. On the night of 5/6 June 1944, the 36th and the RAF efficiently screened out the enemy's early warning radar systems, a massive step in ensuring the success of the D-Day invasion on 6 June.

A month later 91 Group of RAF Bomber Command took possession of Cheddington airfield, returning it to the original role of a training base.

During their service at Cheddington, the US airmen were expected to 'work like hell' on-duty but were free to do what they liked off-duty, subject to regulations. Several buses left the gym at Cheddington every night at 7 p.m. for the American Red Cross Club in Luton, returning at 11 p.m. There were dances in Leighton Buzzard and nearby villages, providing opportunities to meet local women: sixty-seven US airmen got married as a result.

Thomas D. Thompson, an engineer/gunner with the 406th (Night Leaflet) Squadron in 1944, flew forty-five missions from Cheddington. He recalls often cycling into Cheddington village to visit a pub which had a 'Lion' on the sign, where the landlady always bought the 'Yanks' their first drink. The drink had to be gin and orange, as that was her drink. After that, it was mild and bitter. Thompson also remembers taking tinned fruit to an inn on the canal to exchange for a cooked English breakfast. In May 1978 he visited Cheddington again and the sight of the old airfield brought back a lot of good memories. He met many friendly people in Cheddington, Tring and Leighton Buzzard and felt at home

Memorial plaque near the Water Mill at Ford End Farm, Ivinghoe, erected on 15 November 2009, in remembrance of the crash of a B-24 Liberator shortly after take-off from Cheddington airfield, with the loss of two crew members.

there, he says, because his great-grandfather was born in Appleby, Cumbria. The young Thompson left us the following thoughts:

What will become of this little section of England that has been our life for the past two years? Will the runways and hard standings lie abandoned and untouched in mute tribute to the men who worked, lived and sweated out the planes during the years of war and restriction? What will become of the dogs once so well fed round the mess halls – the legion of station mascots? Will they walk through deserted kitchens wondering where the chow lines and their GI pals have gone? What impression have we Americans made in our frequent contacts with the English people who have been our neighbours and companions during these long monotonous war-weary years?

The plaque at Mursley water tower, in memory of the four crew members of a Wellington bomber of No. 26 OTU on 11 April 1943. This was one of many tragic accidents involving No. 26 OTU, a RAF unit formed at Wing in January 1942 for training night-time bombers.

The airfield was transferred to Technical Training Command after the war, but they had no use for it and it closed in February 1948. Part of the airfield was retained for use by the army, until the site was officially closed in 1952. However, press reports in the *Luton Evening Post – Echo* newspaper suggested that it was then used for secret purposes, reputedly a CIA chemical weapons dump for a 'behind the lines' special operation codenamed 'Operation Gladio', to be put in place if the Warsaw Pact countries invaded Western Europe. Local searches after the Great Train Robbery in 1963 revealed the existence of this formerly secret activity, and the BBC used a current affairs programme a few weeks after the robbery to expose the base to the world. A small part of the airfield was occupied by the US military up to 1975.

Despite losing his livelihood, tenant farmer Arthur Reeve ended up as a groundsman looking after the airfield, and became the Americans' number one fan. He continued to live in Church Farm, Marsworth, and the land comprising the airfield was sold off by auction when the Americans finally moved out. Today the runways, hardstands and aprons have all been removed for hardcore. There are various abandoned buildings on what was the technical site along the south-eastern part of the original perimeter road, and a grass runway used by private light aircraft can be seen on Google Earth satellite images.

The definitive account of the *Secret Squadrons of the Eighth*, by Pat Carty, was published in 1990 and contains comprehensive appendices and a selection of rare photographs.

The brick memorial just outside Cheddington airfield. The inscription reads: 'Dedicated in October 1982 to the US 8th Army Air Force stationed here 1942–45.'

9

THE BILLETING NIGHTMARE

The many billets provided by Leighton Buzzard for military and civilian personnel employed at the various establishments described above had a major effect on the existing civilian population. For more than a century, the families and retainers of the well-to-do had been living in substantial houses in, and on the outskirts of, Leighton Buzzard and Linslade. In the late Victorian and Edwardian eras, when the Rothschild banking family entertained the Prince of Wales and the aristocracy to hunting parties at their nearby stately homes, many large houses were built around the town. Between the First and Second World Wars, the hunting waned and with the decrease in the number of people willing to take up a life 'in service', many of these houses had become a burden on their owners. It would have been with some financial relief but personal sadness that they realised the government and the military were eager to requisition many of these houses to accommodate the growing numbers of specialists and service personnel arriving in the two towns. To make matters even more difficult, several army regiments would be posted to the area for training and two groups from the Women's Land Army (WLA) would be stationed in Leighton Buzzard and Linslade and would need somewhere to live.

Not only were buildings needed for administrative purposes, but residential accommodation was also in short supply – as previously mentioned, RAF Leighton Buzzard had no barrack blocks or dining facilities and No. 60 Group's permanent HQ at Oxendon House had no sleeping quarters for those working there. Several houses are mentioned specifically in connection with accommodation for these groups and for Bletchley Park – the increasing importance of code breaking as the war progressed required many more people to be recruited and found homes. As well as the known addresses, personnel were fitted into every available space.

One unfortunate consequence of the requisitioning of the Leighton Buzzard Public Assistance Institution (the old Workhouse) in Grovebury Road on 29 May 1940 was that the inmates had to be relocated to Ely and Doddington in Cambridgeshire. There were pathetic scenes as the elderly were taken away in specially equipped vehicles. Many were locals and had been visited regularly by volunteers whom they had come to regard as friends. They felt the parting very acutely.

Charles Henry Swaffield, who had been Master of the Institution for thirty-eight years, and the Matron, Miss Ethel Levette, who had been there twenty-seven years, personally supervised the move and gave the 'old folks' parting gifts. Rev. A.T. Stephens, a regular and popular visitor to the Institution, and George Chapman, formerly Chairman of the old Board of Guardians, were among those who saw them off. Nine men, the oldest of whom was 85, went to Ely and fourteen women, one aged 95, went to Doddington. The staff of the Institution and two or three of the men remained for a while to help clear up. The ground floor of the building was turned into an Air Raid Precautions (ARP) HQ and a maternity and child welfare clinic; it later housed WAAFs working at RAF Leighton Buzzard.

Many large local houses were converted to accommodation for army personnel during the war. Among the properties requisitioned to meet the needs of RAF Leighton Buzzard and No. 60 Group, in an order dated 8 April 1940, was The Albion, 3 High Street, a former temperance hotel opposite Church Square housing fifty-eight WAAFs. The RAF was not happy with this property, regarding it as:

a typical small old fashioned town hotel with many dark corridors, awkward corners and sundry steps up and down. The rooms are mostly small, though fairly satisfactory for two or three WAAFs. The kitchens and scullery facilities are reasonably good, but the dining-room and recreations rooms are much too small.

'Woodlands' was on the western side of Plantation Road. Initially it housed 152 WAAFs. It was described as 'A large rambling house, situated in its own grounds about 1½ miles outside the town'; in October 1941 forty more WAAFs were housed in huts built on the lawns. 'The situation and aspect are excellent, but the building has too much corridor and too many small rooms for convenient administration … the dining and recreation rooms are too small.' One large recreation room was used as temporary sleeping quarters until ten more barrack huts were built, including a 'latrine and ablution block', which allowed a further 600 WAAFs to be accommodated.

The Grange (formerly known as Heath Cottage) at Heath and Reach, in the Woburn Road, took another 91 WAAFs. In 1934, when it was sold at auction, it was certainly a substantial property 'screened from the road by trees and a high brick wall' and with modern drainage, mains gas and central heating, in addition to which 'Electric Light and Telephone Service were available in the village'. It

The Cedars, Leighton Buzzard.

The Albion Hotel on the right of the picture in Church Square, a temperance hotel that was requisitioned for the use of WAAFs stationed at RAF Leighton Buzzard. It was one of the better places to be billeted.

is known that ninety-one WAAFs were at one time living here, two or three per room, and this is possibly the place referred to by Josephine Pipon in a letter to her parents, written on RAF letter-headed paper and dated 28 November 1939: 'WAAF Hostel, Heath and Reach, Leighton Buzzard, Bedfordshire ... This place isn't bad – quite luxury from West Drayton. No more polishing floors, coping with stoves ... no more marching to meals laden with eating implements and no more queuing for baths in a washhouse.'

Another requisitioned private property, 'Wilmead' in Stoke Road, Linslade, housed forty-six WAAFs. Prior to February 1926, when it was bought by Lady Emily Anna Acland, it had been known as Bossington House. All that remains today is its tree-lined carriageway, now Lime Grove. In 1935 The Heath, a former Bassett family house in Plantation Road, where Heath Court now stands, was

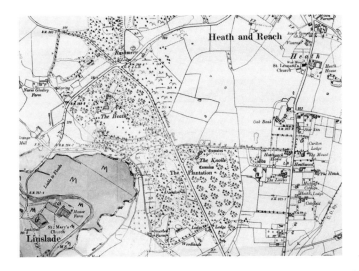

A map of the Plantation Road and Heath Road area of Leighton Buzzard showing some of the big houses that were requisitioned by the military at the beginning of the war. They include Carlton Lodge and Oxendon House, required for RAF 60 Group, and Woodlands and the Heath, which were used as billets and surrounded by Nissen huts to house hundreds of airmen and women from Q Central, Bletchley Park and other units.

owned by William Alfred Lailey Rowland, High Sheriff of Bedfordshire in 1935, who died on 4 February 1940. The house became an annexe to Oxendon House when, once again, the numbers working with No. 60 Group outgrew the premises. Castle House in Linslade accommodated a Royal Artillery (RA) unit; Lake House in Lake Street a section of the King's Royal Rifles; and The Hollies and Vale Stables in Church Road, Linslade, a section of the Auxiliary Territorial Service (ATS). An Army Dental Centre was set up on the upper floor of 41 High Street. Among the conversions for military use were The Knolls (another Bassett House) and Oxendon House, HQ of RAF 60 Group.

Later in the war, the woods between Woburn and Little Brickhill were used as a staging area for D-Day and during the winter of 1943–44 a large army ammunition dump was constructed there. Several live mortar shells were found when a new golf course was laid out there several years ago. Also in these woods was a gas cleansing centre, containing, according to local builder Raymond 'Viv' Willis in his book *Born to Build*, enough shells of gas to poison the whole of the British Isles. The shells were as big as 7ft (2.1m) long and 8in (20cm) wide. He once saw a 3-ton lorry arrive at Little Brickhill with two large leaking cylinders of mustard gas. The officer-in-charge ordered 200 men of the Pioneer Corps to put on protective clothing and to dig out the earth below the lorry to a depth of 20ft (6m), gradually lowering the lorry as they went, until it reached the required depth. The soil was then shovelled back on top of the lorry and the site levelled. As far as is known, the lorry and its load are still there.

THE CHELSEA PENSIONERS

After bomb damage to the Royal Hospital in 1941, the Rothschild family invited a group of Chelsea pensioners to make their home at Ascott House, where they remained until 1947. Several of the pensioners died while at the House:

19 July 1943	John Helstrip, aged 73
15 November 1943	Thomas J. Mason, aged 79
13 January 1944	William Johnson, aged 70 or 81
22 March 1944	William Roser, aged 82
22 August 1944	Alfred Giles, aged 88
13 October 1944	Albert Edward Glazier, aged 76
20 December 1944	Albert Henry Snellgrove, aged 81

Stockgrove Park

At Heath and Reach, one of the first jobs undertaken for the military by the local building firm William George Willis & Son was to convert Stockgrove Park Mansion to house 200 men of the 8th and 10th Companies of the Royal Army Medical Corps (RAMC) under Lieutenant-Colonel Tristram Samuel, a man who had come up through the ranks and expected everything to be spotlessly clean, even to the extent of requiring the latrine buckets be polished. Among the requirements were twenty toilets, with ten in a row. Denis Argent of the Non-Combatant Corps (NCC) records in his book *A Soldier in Bedfordshire 1941–42* that he was employed on digging out old sewage soakage pits for the 'big house'. Apparently the smell was only mildly unpleasant. Argent says that the other work at Stockgrove Park was not too hard and *al fresco* lunches, with breaks for cocoa mid-morning and afternoon round a blazing woodland fire, were very pleasant. On one occasion, a man and woman riding past along the drive in smart riding habits, identified by the NCC as the millionaire owner 'Old Killjoy' and his wife, who were thought to have sacked their own staff at the mansion once they had RAMC/NCC help, got what Argent describes as a 'conchie' reception – a raspberry or two and a loud shout of 'is there a war on?' The landowner riding by was Ferdinand Michael Kroyer-Kielberg, chairman of the United Molasses

Company, who had commissioned the building of the mansion in the late 1920s and moved out to the dower or guesthouse to allow the army to move in. He was born in Denmark and was chairman of the Anglo-Danish Society, which, under his leadership, played an important part during the war in supporting the Free Denmark Movement. He was knighted in 1947. His son, Peter Emil Kroyer, was awarded the Distinguished Flying Cross (DFC) in 1945.

At some point before May 1941 the RAMC moved to The Heath in Plantation Road. Later in the year Stockgrove House was occupied by 228 Battery of the 66th (Lowland) Regiment RA. One artilleryman writes that this:

> must have been about the Army's best billet. 228 Battery dwelt there in considerable luxury especially as Mr Kroyer-Kielberg had stipulated that the oil-fired central heating was to be kept in operation throughout the winter months, though we heard that in a later year the War Department could not get the heating oil and some of the piping was damaged.

By December 1941 commandos were training in Stockgrove Park and in August 1944 the house was requisitioned as additional accommodation urgently needed for Wrens working at Bletchley Park. Also based there from May to July 1944 was 81 British General Hospital.

In the Virtual Library archives hosted by Bedford Borough Council, under the heading 'Stockgrove Park, Kroyer-Kielberg, William Curtis Green', there is an intriguing reference to a secret meeting of generals in a cottage at Stockgrove in 1940–41. The location is identified as 'almost certainly Stockgrove Park for many reasons, particularly as Bletchley Park codebreakers were stationed at Stockgrove House and also at the "Duncombe Arms" in Great Brickhill'. The original note about this meeting, by Grace Howell, does not specify the location, but gives various clues that strongly indicate the 'cottage' as being one of the two South

A 30cwt Bedford truck and an RAMC
ambulance in the stable yard at
Stockgrove Park in 1940.

The Boat and Summer House at the 'Bathing Pool' in Stockgrove
Park, destroyed in a fire in 1963.

The South Lodge/Rushmere entrance to Stockgrove Park, identified
as the setting for a secret wartime meeting of eleven generals.

Lodges at the Rushmere entrance to the Stockgrove estate. From here a carriage
way led to the main house. Grace and her husband, on return from Hong Kong,
had rented one of the lodges for 15s a week.

One day an army officer knocked on the door and said that he wanted to
commandeer the house for a secret conference of eleven generals at 5 p.m. He
seemed quite satisfied that he had 'discovered' the right place at last. Right on
time, he arrived with an escort of police to surround the house and asked the

couple to remain in their dining room until the great men had departed. From there they watched in fascination as the generals arrived from different directions, the lodge being at the meeting of four roads. The last to arrive was described by the army officer as 'the little man' and 'the chief of them all'. At about 5.30 p.m. an imposing general, 6ft 4in (1.93m) tall, walked into the room, said he had been ordered to thank them for accommodating the generals at such short notice and expressed interest in the Chinese objects they had in the house. When the grand persons had all departed, Grace Howell asked the army officer, 'in fear and trembling', what it had all been about. He answered, 'I am so sorry to say that I cannot tell you but this I can say, that you have had here in your house Generals of the highest rank in the British army.' He put his heels sharply together and his hand to his forelock, gave a long salute and was gone. Grace Howell stated that the meeting was held at a time of possible invasion, and that is all we know.

Bletchley Park

In 1939, in addition to the people being brought in to run Q Central and No. 60 Group HQ, several hundred more – academics, military officers, technicians and young society women – reported to a Victorian mansion, Bletchley Park in Buckinghamshire, 6 miles (9km) away. This was the centre for deciphering enemy codes intercepted at Q Central. When normal channels proved insufficient to recruit all the staff required, Churchill asked his Chief-of-Staff, General Hastings Ismay, to ensure Bletchley Park had everything it wanted. He emphasised the priority with his famous phrase 'Action this day'. Bletchley Park, along with No. 60 Group's radar and Q Central's communications and all the forces involved, was essential for victory in the Atlantic and North Africa, and in coordinating the D-Day landings. It also produced the world's first programmable computer and a successful Anglo-American intelligence partnership. At its peak, over 10,000 people worked at Bletchley Park and its associated out-stations, women outnumbering men 3:1.

Leighton Buzzard and nearby villages were involved in providing billets for some of the Bletchley Park personnel. Homes in Aspley Guise providing billets for some of the cryptologists and mathematicians were known as 'Hush Hush Houses'. Their work remained shrouded in the highest level of secrecy until 1974, when the first book on Bletchley Park appeared. Indeed, such was the secrecy surrounding the work that there were at least two 'white feather incidents', when men at the base were accused of cowardice by local women for not joining up.

There were some compensations for the long hours working at Bletchley Park. Asa Briggs (later Baron Briggs), a codebreaker at Bletchley Park from 1943 to 1945, recalls in his book *Secret Days* that the BBC Symphony Orchestra, the BBC Theatre Orchestra and the BBC Singers were evacuated to Bedford in 1941 and stayed there until 1946. They became great local attractions, with staff from Bletchley Park a major part of the audience. Vic Howard of Kempston recalls that Stanford Robinson dropped his baton at one broadcast by the BBC Theatre Orchestra. It fell at Vic's feet in the front row so he stood up, retrieved it and handed it to the great man with a sense of importance welling up inside him. Nor can he forget coming out of E.P. Rose & Sons of High Street, Bedford (now Debenham's) and meeting Sir Adrian Boult going in, spotting Paul Beard, leader of the Symphony Orchestra, and Malcolm Sargent in the street, and coming across Sir Henry Wood in town. Quite a full house.

Dennis Bellamy, also from Kempston, recalls returning to Bedford from leave by train in 1941 and finding himself in the same carriage as Sir Henry Wood and his wife Lady Jesse. He mentions being billeted at one stage in what was evidently a boarding school, since there was a notice on the wall saying 'If you require a mistress during the night then please ring the bell.' The same story is told by National Trust staff at Hughenden Manor about American aircrew at a girls' school at High Wycombe, so it seems to have done the rounds in various guises.

But the best place to visit in Asa Brigg's view was the American Camp at Little Brickhill, 3 miles (4.8km) from Leighton Buzzard. Here, in his early twenties, he ate his first avocado, tasted his first American coffee – far better than the stuff at Bletchley Park – and drank his first bourbon and his first dry martini. He was never the same again, he wrote.

Kenneth Halton, billeted in Leighton Buzzard from 1944 to 1945, has left us a detailed description of his billeting experiences in the WW2 People's War archive:

The reason why I went to Bletchley Park at all was, as I see it, that I was a suitably trained Post Office engineer who happened to be available at the time. The Hostel Manager at Bletchley Park, Mr Griffiths, was responsible for looking after the civilians and had established suitable billets (well, billets anyway) in all the towns and villages within about a 10-mile radius of Bletchley Park. Within days we found ourselves billeted in a decidedly shabby place in Leighton Buzzard which was to be our home until the war ended. We did not have much choice because we were not being paid subsistence allowance and the payment for our billet was made directly by the Hostel Manager. There was far more travelling time and it was not so easy to get to and from work for overtime. We got over one of those problems by bringing our bicycles back with us after our next visits home and subsequently enjoyed some very pleasant rides between Bletchley and Leighton Buzzard. The people we stayed with were kind enough but their house had already been condemned when they moved into it to escape from the Zeppelin raids on London in 1917 and they had therefore done nothing to improve it since.

The house had only one cold water tap, in the kitchen, and no bathroom. There was a toilet – down the garden – but the ravages of time had modified it from a WC into a C. The cistern had long since rusted and ceased to be of any use and its function had been taken over by a bucket of water from the kitchen tap once a day. The principal lighting of the house was from a single electric light bulb in a batten lamp-holder screwed directly into the middle of the living-room ceiling. As the ceiling height was little, if anything, over six feet this would have been decidedly dangerous but for a large circular table which occupied the centre of the small room. There may possibly have been another light in the front room but elsewhere candles were used. The front bedroom had in it a built-in cupboard, large enough to accommodate a child's bed. The cupboard boasted its own small window and the grandchild of the house slept in there.

The tiny winding staircase, in a cupboard in one corner of the living room, led into a landing big enough to accommodate a camp bed, which was used by a Royal Army Signals Corps driver from Bletchley Park. His 'bedroom' was a thoroughfare into the front bedroom and also into the back bedroom which Frank Crofts and I shared. It was like stepping back in time, leaving the fluorescent lights in Block F and returning to our candle-lit bedroom. There was a difficulty too in that we were both studying correspondence courses for our telecommunications examinations but we overcame that problem by buying some wireless-set accumulators and using them to light a row of four flash lamp bulbs in small stands which we made. In modern parlance these were lead-acid secondary cells which we were able to recharge in the workshop and I am pleased to say that we both passed our exams alright.

Kenneth Halton then goes on to describe his impressions of an event that occurred shortly after the end of the war:

Frank Crofts and I were cycling back one sunny day to our billet in Leighton Buzzard and were somewhere near Stoke Hammond when we heard the sound of aircraft – lots of them. Initially a group of fighters came into view, then others, then bombers, more and more of them. These were US planes and, unlike the Lancasters, they were flying in the orderly formations beloved of the US Army Air Corps. The fighters were in neat groups of 19 with three groups in a vee formation of 57. There were Thunderbolts, Mustangs and

Lightnings surrounding a huge array of Flying Fortresses and Liberators. The tidy formations made counting a little easier but the very large number made accuracy impossible. My tally was about 1500 and this was later confirmed in a newspaper report that the US Eighth Air Force – the 'Mighty Eighth' – had made a victory celebration flight over southern England that day.

Jeanne Isaacs, a WAAF from 1942 to 1947, describes in the WW2 People's War archive her experiences as a teleprinter officer at Bletchley Park:

I had no idea how important Bletchley Park was or what was being carried out there. But as soon as I arrived, off I was sent again, to Chigwell in Essex, whilst I was being 'vetted', and at Chigwell we were taught the Morse and Murray codes and how to read them from tapes coming through machines from almost everywhere. Of course this was all before computers etc. After three weeks at Chigwell, I returned to Bletchley Park where I joined a team of teleprinter operators and soon settled in. Our accommodation was in Nissen huts, with on average 35–40 WAAFs in each one. The loos and showers were situated elsewhere and so, if we were on the late shift or nights, it meant us going out in the sometimes pouring rain, or maybe even in snow to spend a penny or to shower.

A few weeks before D-Day, all our leave and weekend passes were cancelled. We guessed something was up. It was – all hell broke loose. I was on duty the night before D-Day, and all the machines went mad – tapes came spilling out

Squadron TPO 'putting on a show' at the Corn Exchange in Leighton Buzzard in 1944. The Corn Exchange building was used as a social centre for RAF and other personnel. Shows were put on by ENSA ('Every Night Something Awful').

… there was [*sic*] barely enough hands to collect it all. These tapes had then to be transmitted to the teleprinters, the results of which were sent down to the bowels of the building to be decoded in the Enigma section.

These hectic days went on for some months and then, eventually, VE Day. I decided to stay in the WAAF, and make a career. I was going overseas, but unfortunately my mother became unwell and I was asked to take over my mother's role in the fish and chip shop. By then it was May 1947, and so I had been in the WAAF for almost five years.

10

A SHARP INTAKE – A TIDE OF EVACUEES

Despite the accommodation problems already created by setting up this group of secret bases, another branch of government, the Department of Health, had apparently independently decided that the town was a key area for the evacuation of thousands of children. In a long-planned exodus from vulnerable areas in 1939, entire schools were packed onto trains and into fleets of coaches and sent into the country.

Although this unprecedented movement of people was remarkably well organised, it must have been a terrible shock to such a small town. At the start of the war Leighton Buzzard had a population of just 7,897. Over the next few months the people of that town, of the adjoining town of Linslade and of the villages immediately surrounding them both would have to absorb several thousand individuals. Local people, often set in their ways, were sharing their homes with strangers of varying backgrounds and unpredictable habits. Almost all of these were from London and many were children; some came with members of their family or community, others alone. All were no doubt anxious about the future, visitors and hosts alike.

It is clear that the authorities that ordered the evacuation had no idea what was happening in Leighton Buzzard because the idea was to withdraw the vulnerable to a place of greater safety, not a place certain to be a German bombing target. Anywhere perceived to be reinforcing the strength of the RAF was a potential focus for enemy air raids. The obvious targets were airfields with reconnaissance aircraft, fighters and bombers, but of equal importance were identifiable industrial areas where aircraft, machinery, weapons and vehicles might be made, along with ports and harbours whose shipping transported these and vital food supplies and personnel. Consequently, the people who lived and worked in these places were in great danger.

Had the authorities behind the evacuation suspected for a moment the key role that Q Central and No. 60 Group were playing in the Battle of Britain and the secret war, they would perhaps have thought twice about placing so many children there. Either they were never told or they gambled on the Germans not finding out. The evacuees and Leighton Buzzard were lucky.

The evacuees' story began in the summer of 1938, when the government asked Sir John Anderson (later Lord Privy Seal) to form a committee to plan for the evacuation of all children from Britain's large cities. Parents were encouraged to make their own arrangements, but provision would be made for those unable to do so for moving groups of schoolchildren in their teachers' charge. To this end, Anderson decided to divide the country into three zones – evacuation, neutral and reception. The evacuation areas were urban districts where heavy bombing raids could be expected and which residents were advised to leave. Places designated neutral were those where evacuees would neither be taken nor sent. Reception areas were 'safe' places where evacuees would be lodged. Residents of reception areas would be legally obliged to take in evacuees.

Advertisements were placed nationwide in local newspapers by the Ministry of Health, to persuade people living in relative safety, well away from cities and ports, to share their homes with the hundreds of thousands of children being sent far from the terror of air raids. These advertisements were intentionally emotive for billeting was to be compulsory, while evacuation itself was not. Those living in rural areas with even the smallest space available on their premises would be forced to take in evacuees. Many complained about it.

Accommodation was to be by billeting in private houses; although the initial cost was to be borne by the government, those able to contribute to their own maintenance would have to do so.

By the summer of 1939 the Home Office had instructed the authorities in Leighton Buzzard and Linslade to find homes for 1,806 evacuees should an emergency arise. Although war was officially declared on 3 September, the order to evacuate was given in a radio broadcast at 11.07 a.m. on Thursday 31 August. Operation Pied Piper began in earnest the following morning.

For those born after 1945, whose knowledge is conditioned by old television and film coverage, mostly in black and white, the word 'evacuee' is likely to conjure up a picture of a railway station and a lone, bemused English child in a belted mackintosh. The child will probably have a luggage label attached and perhaps be carrying a small brown suitcase with a gas mask in its cardboard box slung across a shoulder on a piece of string.

Although the majority of evacuees were, indeed, both English and children and much like this stereotype, there is an abundance of evidence that people of all ages and from many nations, often in groups already known to each other, sought refuge away from homes which for various reasons would be likely prime targets of enemy action. In that first wave, around 1.5 million vulnerable people were moved to safer parts of the country. The statistics derived from national

Leighton Buzzard Urban District Council

Council Offices,

Leighton Buzzard.

REGINALD F. A. TUTT,
BARRISTER-AT-LAW.
CLERK OF THE COUNCIL.

12th October 19 38.

J. B. Graham Esq,
    Clerk of the County Council,
      Shire Hall, BEDFORD.

Dear Mr. Graham,
              Evacuation.

          I thank you for your letter of the 7th inst, and for the expressions of appreciation in connection with the billeting arrangements for evacuees.

          With regard to the experience gained in this area, I have pleasure in answering your various points as follows:-

1.   The whole of the district.

2.   The general response was satisfactory, and I should estimate that 10% of the houses were willing to accommodate children.

3.   No. Leighton Buzzard station is in the urban district of Linslade and about a mile from the centre of Leighton Buzzard.

4.   I am of opinion that there would have been a considerable difficulty in accommodating the 3,100 refugees on the basis of the standard laid down by the Minister of Health. For this reason: It was found that a considerable number of houses in Leighton Buzzard were already occupied, and about to be occupied, by friends and relations of the householders. Also, it was understood that several hundred children were coming here from the Waifs and Strays Society. Assuming that these children were part of the official 3,100 (which appears to be uncertain) I should estimate that at least 500 should be taken from the figure of 3,100 which was originally estimated.

5.   Complete arrangements had been made in this respect, and premises had been earmarked, by arrangement with the Owners, for this purpose. No difficulty of any kind was contemplated.

              REGINALD F.A.TUTT,
                Clerk of the Council.

Council Offices,
    Leighton Buzzard.
      24th September 1938.

Preparations for evacuation of urban children were being made a year in advance of war being declared. The Ministry of Health had already contacted Bedfordshire County Council regarding potential accommodation on 16 September 1938, as shown by this letter from the Clerk of Leighton Buzzard Urban District Council.

registration at the end of September 1939 show that this number comprised 393,700 unaccompanied schoolchildren, 257,000 mothers and accompanied children, 5,600 expectant mothers, 2,440 disabled people and 89,355 teachers and other escorts. In addition, many thousands of wealthier individuals chose to distance themselves independently from danger, some settling in hotels for the

GOVERNMENT FIGURES FOR EVACUEE ACCOMMODATION IN BEDFORDSHIRE AND BUCKINGHAMSHIRE

	Bedfordshire	Buckinghamshire
Total accommodation available February 1939	110,945	116,245
Percentage privately reserved February 1939	11%	27%
Number received September 1939	37,163	31,345
Percentage received to total accommodation	33%	27%
Percentage received to numbers expected	45%	47%

duration and several thousands travelling to Canada, the United States, South Africa, Australia and the Caribbean.

Even though theoretically there was room for all these people, finding them accommodation among a sometimes unwilling population, when thousands of places had already been requisitioned by the military, was another matter. The unenviable task went largely to the Women's Voluntary Service (WVS) – assigned responsibility for the Government Evacuation Schemes by the Ministry of Health in September 1939. Responsibility for organising the reception of evacuees and putting them together with locals was given to Relieving Officers – these had until 1929 been officials of the Poor Law Unions who handled applications from the destitute for assistance in or out of the Workhouse.

The county councils appointed billeting officers, again often members of the WVS, who were detailed to find host families and deliver the evacuees to their new homes. These billeting officers, of whom Leighton Buzzard appointed twenty-five, would oversee District Visitors, who in turn organised Welfare Committees in each village.

Although the rights and powers of the billeting officers were always something of a mystery, this structure was on the whole extremely efficient and absorbed the teachers and children evacuated from London during the Blitz. One schoolboy evacuated to Leighton Buzzard from London in 1939, Ronald Frith, writes:

Schools and communities were to organise the children into groups, baggage ready packed, labels round their necks so that as soon as hostilities commenced, trains could whisk them away from danger as quickly as possible. And so in September 1939, my mother, my brother Derek (aged 2) and myself (aged 6) found ourselves boarding a train at Willesden Junction to head north, we knew not where, to be evacuated to the country. The trains were specially chartered and full of evacuees, some excited at the prospect of new adventures, some, particularly the young ones, bewildered and tearful. Some were sad and fearful for the future for the outlook seemed bleak.

Leighton Buzzard was a major 'detraining centre' for Bedfordshire and on Friday 1 September 1939 800 children arrived on one of the first dedicated evacuee trains, a train so long that those in the front carriages needed to alight before the rear could be drawn forward alongside the platform to disgorge the remainder of its passengers.

This part of the evacuation process proved to be extremely well orchestrated and, during the next three days, more than 7,000 evacuees disembarked at Leighton Buzzard's railway station, many of whom would be taken by bus to other parts of the county. This effectively almost doubled the population of the town temporarily and some indication of the response to this unprecedented event can be seen in the local newspaper account, dated Tuesday 5 September 1939, of the arrival of that first train:

> A crowd of sightseers thronged the entrances to the Station and the Recreation Ground, but they were not allowed by the Special Constables and the regular police to get nearer. The police carried their gas marks and steel helmets. For some time before the train pulled up at the platform the local evacuation officers were waiting in the Station approach, which was closed to the public. Thirteen buses were also waiting to move the children to the villages.
>
> As soon as those destined to remain in this area for the foreseeable future had alighted, they were marched to the recreation ground near St Barnabas church in Linslade, each with a gas mask, a parcel or bag and a name-tag.
>
> The work of sorting out the children was greatly helped by microphones and loud speakers, through which Mr M.C. Clifford [the Superintendent Registrar of Linslade at the time] was able to call up the parties to the medical hut.

Stuart Hammond, from Acland Central School for Boys, Kentish Town, remembered:

> We all got off the train and our teachers directed us to a grassy bank outside the station to await allocation to foster parents. We were all given a large bar of Cadbury's chocolate! I was eventually ushered with eleven other boys onto a coach and we were driven a few miles to the village of Great Billington.
>
> The Leighton Buzzard party were first taken to the Temperance Hall and provided with a meal. They were then taken back to the Town Hall and placed

Evacuees leave Leighton Buzzard station, some with their mothers, others helped by volunteers. They were gathered in various public halls and most were found billets within a few hours.

in the charge of the billeting officers, who distributed [them] to homes as arranged during the previous house-to-house canvass.

Collecting centres were The Corn Exchange and The Temperance Hall, Lake Street; the classroom at the rear of the Council Offices, North Street; the Church Rooms, Church Street; the 'Old Vic' Concert Room, Hartwell Grove and The Forster Institute, Waterloo Road, Linslade. In most cases in Leighton Buzzard, evacuees were allotted to specific billets. There would always be some, though, for whom 'permanent' accommodation had not been found before their arrival.

Beryl Knox and Iris Dewey from Hackney 'stayed in a large hall for a few nights with a lot of other people from London, plus a German spy who was arrested by four men in long black coats and Trilby hats'. Ronald Frith from Willesden recalled:

The train drew into Leighton Buzzard station and we were instructed to get out. We had some refreshments and the nurses examined everyone for head lice, fleas etc. Then we boarded buses to be sent to our destinations in the surrounding villages. It was thus that we ended up in Heath and Reach, billeted with Mr and Mrs Eaton of Lanes End. It was obligatory for householders with spare rooms to take in evacuees and so neither party had much say in the matter. However,

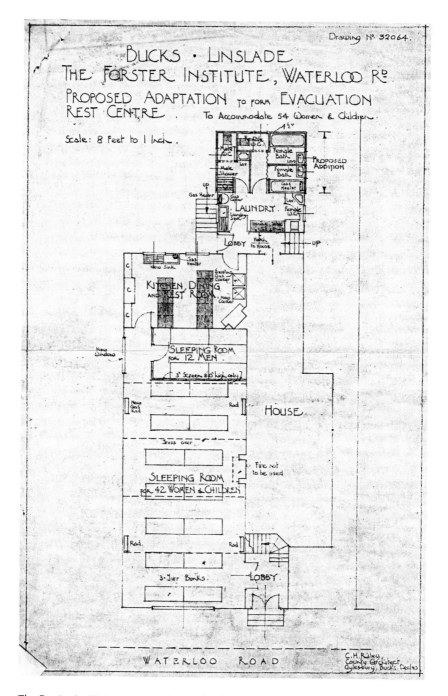

The Forster Institute was one of many local buildings to be converted for 'war' use. The increase in population meant that all accommodation was at a premium and the two rooms were to be sub-divided to allow for its use as an evacuation rest and feeding centre.

Letter to the
Surveyor, Linslade
Urban District
Council from W.G.
Brown, Acting
County Architect of
Buckinghamshire
County Council.

Bucks County Council.

TELEPHONE N° 370-6 AYLESBURY.

ARCHITECTS DEPARTMENT.

County Offices,
Walton Street.
Aylesbury.

Enclosure s

EFRS-H/W.

W. G. BROWN.

ACTING COUNTY ARCHITECT.

27th May, 1941.

Dear Sir,
              Forster Institute, Linslade.

Herewith I send you a diagrammatic plan of the proposed
adaptation of the above Building as an Emergency Rest &
Feeding Centre.

Owing to the urgency of the matter and of the many
Buildings which I have to adapt throughout the County it is
impossible in the time allowed and with a much depleted staff
to prepare working drawings and specifications and to obtain
tenders in the usual way. I am therefore carrying out the
work by Prime Cost Contract under the terms laid down by the
Ministry of Home Security for Emergency Work, and giving
verbal instructions to a selected Contractor.

The plan has been approved by the Ministry of Health's
Architect.

I trust that under the circumstances you will not require
further plans, and that you will permit the work to proceed.

                              Yours faithfully,

                              W. G. Brown

The Surveyor,
    Linslade Urban District Council
    15 Bridge Road,
       LEIGHTON BUZZARD,
       Beds.

we were made welcome and felt a bit more at home when we found out that
Mrs Eaton hailed from Shepherds Bush, not far from our home in Willesden.

Another train full of evacuees, mostly from Willesden, was due on that first day
but only fifty-four of its passengers remained in Leighton Buzzard:

The second train was timed to reach Leighton about 2.30 p.m., but it did
not arrive until 3.45 p.m., and brought much younger children, some of
them under five years old. With their bundles they were a pathetic sight, and
everyone rushed to help to carry them to the Recreation Ground. They stared
wonderingly around all the time, and seemed to be rather bewildered. All kinds
of parcels were carried by these small evacuees. Some had kitbags, others parcels
and some bags that looked suspiciously like pillow slips. All appeared to be very
tired. When they arrived in the Recreation Ground they sat down and did not
move until they had to undergo a medical inspection and climb into the buses.

Although their parents and schools were obviously primed for the event, the children themselves were sometimes given no prior warning of their evacuation. It was hoped that the whole procedure would run more smoothly if no cause were given for premature homesickness. Since the train journeys might last many hours and refreshments might not necessarily be available, some of the luckier children set off with sandwiches, nuts and raisins, biscuits, barley sugar or an apple.

William Smith of Hackney, aged 8 at the time, tells us of his experiences on the day he left his home heading, unbeknownst to him, for Bedfordshire:

> There were no lessons given to us, what happened was this: on our first day at Crondall Street [School] they gave every kid a brown carrier bag with a banana an orange and a bar of Cadburys chocolate. We were told to put the bag in our school desks along with our gas masks, which was in a cardboard box with string, which we used to carry over our shoulder. We had to go to school every morning and stand at our desks. A bell would sound and we had to pick up our carrier bag and our gas mask and then go to the side of the classroom and form up in twos. They would then walk us out into the playground where we would line up with the other classes, the teachers would walk us once around the playground and then take us back to the classroom where we had to put our carrier bag and gas mask back into our desk and then go home. Well this went on for a couple of weeks then one day as we walked around the playground someone opened the gate and we all walked out into the street, there on the other side of the road all the kids' mums were lined up crying and waving goodbye to us, my sister Maudie was crying. I really did not know what was happening, as we went along the road we saw my uncle John. He had gone to the paper shop and bought every comic there and was giving them to the kids as they walked past, nobody had told me or my brother what was going to happen. We found out later that my mum had told my sister Maudie that we were being evacuated and to look after us that was why she was crying. We were walked to another school where we joined up with another crowd of kids. Someone tied labels to our coats with our name and address on then buses took us to St Pancras Station. We all lined up on the platform, a train came in and we were told to get on and someone told us we could eat our fruit and chocolate, when we opened the bag the chocolate had melted and the fruit was all squashie (but we still ate it). There were no 'sell by' dates then.

The government recommended that, in addition to their gas mask and identity card, every evacuee should take the following items: overcoat or mackintosh, boots or shoes, plimsolls, wellington boots, nightwear, comb, towel, soap, facecloth, toothbrush. Boys were to take a vest, a pair of pants, a pair of trousers, two pairs of socks, handkerchiefs, a pullover or jersey; girls should have a vest, a pair of knickers, a petticoat, two pairs of socks or stockings, handkerchiefs, a blouse, a cardigan.

Roy Shaw, who came with his fellow pupils from William Ellis School, Highgate, remembered 'carrying my gas mask and a bag with all my worldly possessions in it'. It would appear that 8-year-old William Smith might have had only the clothes in which he stood up, since he makes no mention of any other luggage. Many of the poorer children had inadequate footwear and clothing, let alone anything suitable for the country lives they were to begin. There was obviously going to be a need for children's and babies' clothing and for the loan of prams and pushchairs. The children arrived in late summer and, in addition to their normal growth rate, there would soon be a change of season. An official report said:

> The wintry weather is increasing the problem of finding clothing for the poorer London children. In spite of all that kindly householders are doing to help them by remaking old clothes, many of the children are still wearing the thin cotton dresses in which they came and have no underclothing. Footwear is also inadequate. There is much criticism of parents who spend money in visiting their children at weekends and bringing them foolish gifts, but who do nothing to see they are properly clothed. One badly clothed family of children have received a letter from their parents announcing that they are having a good time and have bought a new gramophone.

Christina Scott from Dunstable recalled: 'We were summoned back to school early at the Cedars in Leighton Buzzard and had a whole week to collect clothes and underwear to send to evacuees.' The beginning of the school term in Leighton Buzzard and Linslade had been delayed until 15 September 1939 to allow time for the new intake to settle into their accommodation before joining their new schoolmates. By 19 September the *Leighton Buzzard Observer* was reporting that 'Gifts of old clothes, both for women and children, have been generous but the demand is greater than the supply … There is also great need for a sewing-machine.' Bedding, too, was now in short supply:

> On Friday morning Mr Walter Samuel Higgs, Chairman of Leighton Urban District Council, commandeered all bedding and blankets from local shops, but there was still urgent need for more and householders were asked to lend any spare blankets. Invoices had been received from the Government for 80 blankets in transit, but up to Friday none had arrived.

As well as practical help, the need for understanding and support through great emotional strain had always been acknowledged; this was in a climate of government campaigns alerting people to the possibility of enemy infiltration with slogans such as 'Be like Dad, keep Mum' and 'Careless Talk Costs Lives', ensuring that everyone was wary of strangers.

There was a great deal of official concern about the welfare of the children and the reception they would receive. Leighton Buzzard certainly took this responsibility seriously; however, some of the older children, having been tasked by their parents with keeping siblings together at all costs, found themselves unable to keep this promise upon arrival. Though it was in no way their fault, they sometimes had to watch helplessly as small brothers and sisters were taken out of their care. At least in a town the size of Leighton Buzzard it was likely that they would have known each other's whereabouts.

The expectations – and the reality

Prior to the evacuation scheme, those who lived in relatively unindustrialised parts of the country had little idea that conditions in some of the poorer inner-city areas had hardly changed since Charles Dickens and hosts were sometimes bemused by evacuees' unexpected behaviour. One billeting officer reported:

> In a village near Leighton Buzzard some young evacuees were shown into a bedroom the first night and told to go to bed. Half an hour later the housewife was astonished to find the room empty, and was shocked to find the children lying on the floor under the bed. They explained that at home Dad and Mum slept in the bed and they slept underneath it.

Minor crime was the norm for some of the young newcomers: 'My mother pinched this shirt' boasted one juvenile evacuee in Leighton Buzzard as he displayed the garment, 'and my father is the best pick-pocket in London.'

As far as the behaviour of the adult evacuees was concerned, one local licensee staunchly defended his new regular customers. Asked whether there was any truth in the rumours that evacuated women were getting 'under the influence' two or three times a week, he strongly refuted the insinuation: 'They are naturally noisy and think a pub is the best place to work off any exuberance of spirits.'

There must have been stark contrasts between city life and that in the villages around Leighton Buzzard. One billeting report said some of the evacuees in the Wing Rural area declared that they had been 'prostrated by the absolute quiet and absence of amusements' and made mutual arrangements to transfer into Leighton and Linslade where the larger population gave them a greater sense of security.

Leighton Buzzard residents were sometimes appalled at the apparent lack of interest in supporting children, emotionally or financially, of parents who had remained in London: 'Although the parents of some Leighton Buzzard evacuees show great regard for their children's welfare and visit them sometimes more often than they are welcome, there are cases to which the proverb "Out of sight, out of mind" could be applied.' One householder was indignant:

> I am beginning to have serious misgivings about the parents of my little party. They have not sent the children a word – not even a postcard – since they came. The children are well grown with healthy appetites and I am having a job to feed them on the billeting allowance, without counting anything for washing, cleaning and damage to furniture and bed linen. The children need more clothes and new boots. I cannot afford to provide them, and I am beginning to think it is time some sharp measures were taken to remind the parents that, now they are relieved of the cost of keeping their children, they are better able to provide boots and clothing.

Another was overwhelmed by the generosity of her guests' parents 'offering to pay something in addition to the Government allowance because their daughters had large appetites'.

The increase in population had one effect which today might not be as noticeable: some places of worship proved unable to cope with increased congregations. Theodor H. Gaster wrote of 1942 in *The American Jewish Year Book*: 'a Jewish cinema proprietor in Leighton Buzzard, Bedfordshire … placed his premises at the disposal of a local Catholic church, when its edifice had become inadequate through the influx of refugees.' All Saints' church augmented its congregation by at least two choristers, Clifford Gould and Cyril Leeke.

Lack of real understanding on the part of city children about the sources of their food could be a problem – one evacuee, Peter St John, apparently felt such sympathy for pigs about to be slaughtered that he set them free to run down the High Street. To a lot of London children, food was bought, not grown: 'Some of the evacuated children are not accustomed to seeing apples growing in open gardens and are having to learn that other people's property must be respected. A small group were seen stripping an apple tree in Linslade on Sunday morning.'

Wherever they grew up, children in those days were used to playing outside unsupervised, but in Heath and Reach some encountered new sources of danger. Ronald Frith:

> Joining up with our mates, we decided to look for souvenirs (shrapnel etc.) in the sand pit and to see what effect the bomb had on the pit. On entering the sand pit site we spread out, running and shouting as boys do. Part of the pit had steep sides and (my cousin) Oliver was near the foot of such a side. I was away from there when I heard a dull flop and saw that a section of the sand face had fallen – Oliver was buried by about eight tons of sand and I could not see him.

We called but no answer came so we ran back to the house where workmen and Les Arnold were clearing up and making repairs. I said 'Oliver – covered in sand pit.' My mother didn't comprehend what I was saying but Les knew only too well the dangers of loose sand in the pits. He grabbed a shovel, indicated to the others to do the same and ran to the pit. Ten minutes of frantic digging eventually uncovered my buried cousin.

He was nearly asphyxiated. He was of a blue colour and his leg was broken. It so happened that the Chief Constable of Bedfordshire was in the vicinity inspecting the bomb damage. He and another policeman ran to the sand pit and started applying artificial respiration. Mr Eaton, with whom we used to lodge, heard the commotion from his garden, which bordered the sand pit and he scrambled down the sand face to assist. Being a St. John Ambulance man he soon got Oliver breathing again and then they carried him back to the house. The doctor said he would be alright and just needed his leg to be set in splints.

Risk-taking was not confined to the newcomers: John Brommage, a fourth-former from William Ellis, saved the life of a boy by applying artificial respiration after pulling him from the water of Spinney Pool in the converted open-air sandpit. He was awarded the Gilt Cross by the Chief Scout in Bedford for this act of courage.

The people who took evacuees into their homes often complained about the state of their health. Although all were supposed to receive a pre-evacuation medical assessment, the shortage of available medical officers meant that many arrived at their billets unchecked; they sometimes brought additional lodgers in the shape of fleas or head-lice. It was estimated that about 5 per cent of the evacuees lacked proper toilet training, regarding floors and carpets as suitable places upon which to relieve themselves.

There was also psychological illness: many children did not hear from their parents at all and even where there was communication, the stress of separation would often cause bedwetting. This was so widespread that it required official advice: 'Some children may try your patience by wetting their beds, but do not scold or punish; as this will only make matters worse.'

The constant washing of bedding soon represented a serious burden on many hosts and by June 1940 an allowance of around 3s 6d (17½p) a week was approved for householders with enuretic evacuees. This small supplement was added to the government's meagre billeting allowance. A host received 10s 6d (52½p) for taking one child and a further 8s 6d (42½p) for each additional guest, but this was frequently not enough to cover even the cost of feeding the children, let alone replacing outgrown clothing and footwear. Families – children accompanied by their mothers – were provided only with lodging at a cost of 5s (25p) per adult and 3*s* (15p) per child. The fact that the mother was to arrange the buying and cooking of her own food was a frequent cause of conflict when kitchens were shared.

Billeting officials faced many difficulties:

> Since the billeting scheme was drawn up in February there have been many changes, which were not notified. Families have left the town; the accommodation available has altered, and people who were willing to receive evacuees in February have now got relatives in their spare rooms.
>
> Billeting officers who thought their task would be completed once they had found homes for evacuees have been sadly disillusioned. Through a variety of causes, they are still trying to find alternative accommodation for dissatisfied women and children, or to dispose of evacuees billeted on disgruntled householders. It was too much to expect that the task of finding temporary homes for about 1,000 women and children would pass off without a hitch, and it is testimony to the tact and resourcefulness of the billeting officers that the job has been done in most cases satisfactorily. Many of the complaints that are daily coming to hand from both sides are exasperating. Householders and evacuees do not always realise that these voluntary helpers have done all in their power to make each and every one happy and that it is now their turn to pull their weight.

The majority of conflicts during the years of evacuation appear to have involved tension between the resident housewives and the mothers billeted on them. Tribunals were appointed to deal with these cases. Class differences between evacuees and their hosts were often blamed: 'some like roast beef and some like fish and chips, and those two orders of society cannot be mixed. Any attempt to do so means misery for both.' One woman was so full of complaints, many of them trivial, that an exasperated billeting officer offered to pay her fare back to London.

Although some people were evacuated because they had lost everything they owned, others had moved away before disaster struck. On 19 September 1939, the *Leighton Buzzard Observer* reported that many of the evacuees, realising that there had been no air raids on the capital and feeling uneasy in their new rural surroundings, had decided to return home to London, 'even though they have been refused travel vouchers':

> Many adult evacuees have come to regard the elaborate arrangements made for their safety as farcical in view of the lack of 'incidents'. While a large number are thoroughly enjoying themselves in the delightful counties of Beds and Bucks, others are yearning for what is to them the more congenial life in London.

This could not, of course, have been foreseen by the Ministry of Health, which urged every effort to prevent such returns. Even so, eighty women returned home to London, some feeling the need to look after their husbands still working in the capital.

Tensions between hosts and visitors remained. More facilities were desperately needed to accommodate evacuee mothers and young children during the daytime, allowing their hosts the privacy of their own homes for at least a few hours each day:

> As soon as the evacuees came into this district, the WVS opened a centre at the Temperance Hall, Lake Street … This centre has considerably relieved the pressure on the Urban Council officials and has made meetings of the appeals tribunal unnecessary. A meeting was called on Monday week to start an evacuees' club and the attendance of mothers was very encouraging. The trustees of the Hockliffe Street Baptist Chapel have set aside rooms for Club meetings on Monday, Tuesday and Thursday afternoons, and another room as a crèche for young children. The three gatherings held last week were well attended. It is proposed to affiliate the meetings to the British Red Cross Society so that articles may be made for distribution among the fighting Forces. This should materially help women evacuees during the long winter months ahead.

Some of the boys in their teens were understandably always hungry, particularly if they were engaged in agricultural labour, and their landladies could not afford to feed them. Fred Messer, MP for Tottenham South, asked the Minister of Health whether 'he has considered the report of the situation of the evacuees at Leighton Buzzard; whether he can state the number of boys who are to be turned out because the people cannot afford to keep them; and whether he is considering any increase in the billeting allowance?'

The countryside initially offered supplementary free food, unobtainable in the city. Ronald Frith of Willesden said of his host:

> Cycling on his way to work, he would lay some snares and on the way home in the evening take these up along with any rabbits that had been caught. We had rabbit to eat every other day and I liked it … there were plenty of apples. The hedgerows were laden with blackberries and we often had blackberry pie or tart and loads of jam.

Stuart Hammond: 'The weather was sunny and warm and being early September, the plums were ripening and … we ate a lot of these sweet and tasty Victoria's. Unfortunately many of us had stomach ache through over-eating.'

Mass Observation reports repeatedly emphasised the differences between the chronically poor slum-dwellers and the more prosperous rural families upon whom they were billeted, but many evacuees found more primitive conditions with their host families than they had previously enjoyed at home. Henry Gibbins was seven-and-a-half and had lived with his large family in a new council flat in Hoxton, London:

> I was evacuated with my older brother, Billy and two older sisters, Winnie and Ivy, to Leighton Buzzard … All four of us were billeted in a farm near the town. It was very strange to us because the farmhouse had no proper toilets and only oil lamps for lighting. We did not really fit in there and soon we found a new billet.

They certainly moved upmarket:

> Our new home looked very grand, it was a mansion called Mentmore Hall and was owned by Lord Rosebery. However, we had rooms in the servants' quarters and these were tiny rooms that were reached by narrow winding stairs. We never saw the owner of the property and an estate manager and his wife were in charge of the house. He was a friendly man but his wife was not so friendly. I was often sent to bed with no food for coming home late from school. Our parents came to visit us and soon realised that we were not welcome there. As there had been no bombing in London at that time, we were taken back.

Not all the arrivals were strangers to their hosts. Doreen Adcock came with her sister, mother and grandmother to stay with relatives in Leighton Buzzard while her father found them 'a new place to live' after their home in Fulham was bombed. James Toner was billeted with his great-great-aunt:

> I lived in a row of four cottages (in Plantation Road). Cold water was on tap in the kitchen but there was no bathroom. The WC was in a wooden shed at the rear of the house. We had gas lighting on the ground floor but oil lamps had to be used in the bedrooms. There was a gas cooking stove but a wood or coal fire warmed the living room as did a huge coloured portrait of Queen Victoria commemorating her Diamond Jubilee [1898]. We considered ourselves to be comfortable.

Some evacuees arrived as established community groups. Leighton Buzzard was expecting a contingent of schoolboys from the William Ellis School, Highgate some weeks before its actual arrival. Such was the confusion that it was initially thought that a Leicestershire school was being evacuated:

The long expected party of boys, teachers and helpers from Oakham, Rutland, arrived on Friday [13 October 1939], coming into the town on nine specially chartered motor coaches. The boys number 276 and there are about 35 masters and helpers. All come from the William Ellis Secondary School, Highgate, London. Six weeks ago, when general evacuation began, the boys entrained in London for the country, but, as a master says, were obviously put on the wrong train, for they had not expected to go to Rutland. They were billeted in seven villages and Oakham, and although plenty of accommodation was available in each centre, it was found impossible to carry on the lads' education systematically. They were not billeted in forms and the masters found themselves teaching five or six subjects a day instead of the one they were accustomed to. Requests were made to transfer the boys to a larger centre, where better school facilities would be available. Curiously, the Leighton billeting officers were informed nearly three weeks ago that they were to come here, but the masters and boys were certain, up to the last few days, that they were going to Dunstable. The delay in their arrival was due, a master states, to transport difficulties which were a matter for the London County Council and not for the local billeting officers. The boy's ages range from 11 to 17 years and they arrived in Leighton Buzzard complete with luggage, gas masks, etc., and were given a warm welcome by the billeting officers at the Temperance Hall. Accommodation had been found for them more than a week ago, and the boys, masters and helpers were speedily allotted to billets, about 100 going to Linslade.

However, after this exemplary beginning, the absorption of the Highgate evacuees into the local community began to falter, through no fault of the householders supplying accommodation. There was a shaky start at their host school: no facilities were made available to the boys until the second half of term and the situation deteriorated even further. An article headed 'Snob School Protest' was printed in the *Sunday Pictorial* in 1943:

Ninety London secondary schoolboys, evacuated from the William Ellis School, Highgate, to Leighton Buzzard, Bedfordshire have been deprived of facilities at a school, 'The Cedars', to which they were attached – and snobbery is blamed for their plight. The London boys, 400 of them, have shared the classrooms of The Cedars since 1939. Now there are only ninety Londoners left, and they have been told they cannot go into The Cedars except one day a week to use laboratories. They must all crowd into one room at a local Baptist chapel. Mr George Corfield of Hockliffe Street, who has two William Ellis boys billeted in his home, told the *Sunday Pictorial* yesterday: 'First the Londoners were not allowed to use the gymnasium, then they were denied the use of the swimming-pool. They have never been permitted to play on the extensive fields belonging to The Cedars. I have two of these London boys billeted on me, and like many others who give up our homes to make them happy, I am distressed that they should be made to feel so unwanted.' One of the boys said: 'Really I'm not sorry to be away from the place. At The Cedars we always felt we weren't wanted.'

Mr Fred Fairbrother, headmaster of The Cedars replied 'We are losing our traditions with them being here. We can't have meetings of the Science Society or dramatics or things like that which make up the life of a school.' He denied that the William Ellis boys had not had a square deal. On the question of playing fields, he said that since the William Ellis boys played Soccer and his boys played Rugby, it was impossible for them to use the same ground.

This account of his school's merger with The Cedars came from Kenneth Ives, one of the Londoners: 'Throughout the war the two schools were more or less at daggers drawn – certainly the two headmasters were. The headmaster of The Cedars I think adopted the principle that there was not enough room in Leighton Buzzard for two grammar schools and that was that.'

On this occasion, Beaudesert School welcomed them and lent their facilities, including their woodwork master, Thomas Pacey. Mr Pacey was a fascination to the William Ellis boys, according to Kenneth, because 'He had one wooden leg. He would be using drawing pins and would say "I do not need that drawing-pin" and he would stick it in his wooden leg.'

The Highgate school was scattered all over the town during the war years: the main classes were held in the hall at the Baptist chapel in Hockliffe Street; engineering drawing was taught in the Manse belonging to the Primitive Methodist chapel in North Street; books and equipment were stored in an outbuilding at Barclays Bank; the school library was in a barn behind the Black Lion in High Street; the playground was the Town Recreation Ground, between Bell Close and The Sun public house at the corner of Lake Street and Grove Road; the recreation centres were in the Forster Institute, Purrett's Music Rooms, Bridge Street and the Co-operative Society Hall – and once a week the whole school was paraded to the Civil Defence Centre in Grovebury Road to use the showers.

William Ellis being a secondary school, many pupils attained leaving age during the war years; of those original 270, only 148 were still in Leighton Buzzard by October 1940, and another ninety had left by the end of July 1945. Despite the tensions with The Cedars, the experiences of many evacuees were good – they were generally valued in the community and eight of the William Ellis boys met their future wives during their time in Leighton Buzzard.

Three weeks after the evacuation, the *Leighton Buzzard Observer* quoted many of the new residents as being very happy with the town. A lady from Cricklewood said:

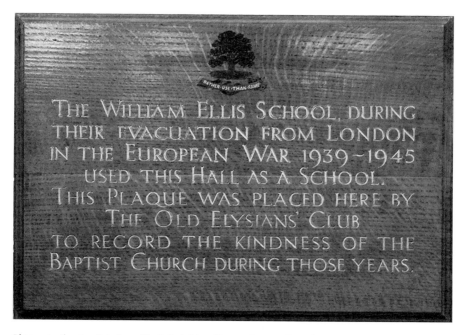

Plaque in the Baptist chapel hall, Leighton Buzzard, recording the wartime relationship between the chapel and evacuees from the William Ellis school.

I shall stay here until the end of the war. I really like the place. A few steps and one is in the country, and what beautiful walks there are. To some of us who never see green fields and animals browsing, it is very lovely. It is also delightful to be able to pick vegetables in the garden; they are so fresh and make such a difference to the meals. I have already noticed the difference in the children. How fine and hearty they are looking.

As stated earlier, not everyone agreed and many went home, only to be evacuated a second time. There were no extensive bombing raids on Britain in those first few months and the phrase 'The Phoney War' began to be used. Many children were back in their homes in the cities and ports by early 1940. The government launched a poster campaign to persuade people to stay put, but with limited success.

The future looked somewhat bleaker, however, after the rescue of over 330,000 troops from Dunkirk between 27 May and 4 June 1940. With the onset of heavy German bombing raids in the autumn, the Blitz, a second wave of 100,000 children were evacuated or re-evacuated.

Beryl Knox and Iris Dewey recalled:

We lived in Hackney, London until September 1940. A landmine fell into the middle of our road and flattened all the houses. Luckily we had gone to the Salvation Army shelter that night, we could hear the landmine but didn't realise it had turned our home into rubble. We lost everything we had. Our parents decided to take us to Leighton Buzzard. We found a small cottage to live in, by a pub, mum, dad and us four children. All we had was a mattress, a penknife and a three-legged table (it wasn't meant to be that way). The cottage had gas light downstairs, oil lamps in the bedrooms, and had an open fire. It was covered in cobwebs with cockroaches in the coal cupboard. Dad worked on the electricity pylons. We lived close to a prisoner-of-war camp for Italian prisoners. After a local river flooded it was the Italian prisoners who came to dig it out. There were no hard feelings between us – the prisoners were friendly and some spoke English. They made willow baskets to sell, and I remember my mother making tea for them. We all felt sorry for them – they did not want to be there. We stayed there until November 1946, we would have happily stayed there, but mum and dad wanted to go back to London. They were eager to see their friends again.

A final mass exodus came in 1944, when Britain came under attack from V1 Flying Bombs and V2 rockets. One million women, children, elderly and disabled people were evacuated from the capital and returning to London was not officially approved until June 1945. By the end of the Second World War around 3.5 million people, mainly children, had experienced evacuation.

Some families settled in Leighton Buzzard who, were it not for the war, would never have known of the town, and the town and its people made such an impression on one expectant mother evacuated there that she named her son in its honour.

PART TWO

THE HOME FRONT – BRAS, BALLOONS, FLARES AND FUN

11

GROWING UP IN 1939–45: PERSONAL MEMOIRS OF BILL MARSHALL, HANS-WOLFGANG DANZIGER AND TONY PANTLING

While Leighton Buzzard became increasingly important as the hub of the secret war, the town itself did exactly what the posters asked – kept calm and carried on. Apart from not having much alternative, the population was also unaware of the vital war work that was going on around them. Q Central in Stanbridge Road and 60 Group based in the big houses along Plantation Road appeared to be just two of the many groups of servicemen and women based in the town.

Like other places that were seen as safe havens, the area also attracted evacuee businesses from London. Some were simply fleeing the Blitz, while others took over existing factories and made different products needed for the war effort. Others were small new factories making specialised items safe from German raids.

In the following few chapters there are snapshots of life in the town. In between stories about the undergarment factory that turned out barrage balloons and the cosmetic firm that made flares are the personal memories of some of the people still alive who describe what it was like living in Leighton Buzzard and Linslade in the war. These were youngsters growing up among the excitement and fear of a conflict they hardly understood while their parents remembered only too clearly the horrors of the First World War barely twenty years before. With thousands of young people in uniform serving their country but not knowing what lay in store for them, there was also a vibrant social life – dances, cinema and entertainment every night of the week.

Alongside the secret war, there were the efforts of the Home Guard, Women's Land Army (WLA), Women's Voluntary Service (WVS), National Fire Service (NFS) and Air Raid Precautions (ARP) wardens. This was the local community war effort carried on in almost every town. Perhaps the WVS did more practical work to help the community than most. For example, at their HQ in Grovebury

Road they had quiet rooms for the forces, including facilities for a bath. In three-and-a-half years they recorded 3,618 baths.

There were frequent money-raising events to buy a Spitfire, a warship and a tank, plus a great deal of effort to salvage metals and paper. All this was vital for keeping up discipline and morale and allowing everybody to be seen to be doing their bit as part of a team. Among the individual efforts were writing to the troops to boost morale and sending them snapshots from home, to keep them in touch with what their growing children looked like during their long absence.

Newspaper cuttings throw light on this strange world and school logbooks indicate teachers' struggles to educate local and evacuee youngsters, but the memories of local residents, who recall some curious episodes and impressions not found in the record books, are perhaps the most entertaining source for what the town was really like. It is amazing to discover the different things people remember – from dates with posh female evacuees to where the machine-gun posts were located. We start with two sets of contrasting recollections of local men, Bill Marshall and Tony Pantling, who were boys when war was declared, and a related third account by a Jewish refugee from Nazi Germany, Hans-Wolfgang Danziger.

Bill Marshall

I was 10 years old at the outbreak of war in September 1939 and have clear memories of that time. My sisters and I were on holiday from Kent – staying with an aunt and uncle in Wing Road, Linslade during August and I clearly remember all the talk and anxiety among the adults at the impending threats of war. In view of these, my parents decided that it would be safer for us children to stay in rural Buckinghamshire rather than in a Kentish suburb of London for obvious reasons. And so I stayed in Linslade for the next six years until the Japanese surrender at the end of the war.

My early recollections … of those late days of summer 1939 were of excitement at the forthcoming conflict. For me and other boys the talk was of battles and victories won and enemies defeated – sooner rather than later. Like Figaro we thought 'What a glorious thing is war' not seeing at our age the darker and sadder aspects of a world conflict. More realistic and chastened views were to come later after a succession of defeats in battles on land and dreadful losses at sea – particularly in merchant shipping. I recall at that time most people – adults as well as children – were confident that the war would be over in months, little realising how over optimistic we were. Thus it was that we entered into war, young and old alike not realising just what lay ahead.

One of the early events of the war was the arrival in Linslade of hundreds of evacuees from London. Soon we were meeting boys and girls new to Linslade and to our schools. Billeting officers visited almost every house in the town to arrange for children to be accommodated. As far as I recall most people responded except of course for the very elderly or incapacitated who were unable to cope with children in their homes. In addition one boys' school – namely the William Ellis School – eventually landed in Linslade and became noticeable by its individual school uniform.

There was also a small group of about twenty or so German/Jewish refugee children who came to Linslade. I believe that they came here through the good offices of a well known London magistrate Mr (later Sir) Basil Henriques – a friend of Mr Anthony de Rothschild of Ascott House, Wing. The children were accommodated in the stables complex of Southcourt Stud where they were presumably cared for by staff employed by their benefactors. I knew all of the boys very well and remember being amazed at how talented and clever they all were. A few names I remember – Peter Laufer, Ernst Goldman, and Franz (or Hans) Oppenheimer. They all spoke perfect English and mixed well with us English boys. They left us after a year or so and I never heard of them again. One other German boy – a refugee I think – by the name of Hans Danziger turned up in Linslade and lived in Woburn House on the corner of Waterloo and Victoria Road and was cared for by a lady I think was called Twissy Clark – what became of him, I know not. It was remarkable that the existing schools managed to accommodate all these additional children at such short notice.

Other newcomers to the town were many soldiers. Much of Church Road was taken over by them including The Hollies, all the stables and bothies that ran from opposite The Hunt Hotel round to the former Clarendon Hotel. They also occupied stables in Church Road now converted into houses presently known as Rochester Mews. The troops were mainly from The King's Royal Rifle Corps – part of the Light Infantry. I believe that they were support troops attached to RAF Leighton Buzzard. The building on the corner of Church Road and Wing Road was a military field hospital. It had previously been a coal merchant's shop, George Chandlers, also selling animal foods etc.

There were also airmen and women in Soulbury Road. Milebush House – a large country residence was taken over and occupied and alongside a camp of timber huts was erected to house mainly Auxiliary Air Force women. These timber huts were subsequently used to house German prisoners when the war ended – where they stayed until such time as they were repatriated … During this latter period they worked locally on farms and building sites. The surrounding villages also added to the number of service men and women seen in Linslade and Leighton Buzzard. RAF Leighton Buzzard came into being and also Cheddington (American) and Wing (RAF) airfields, bringing in several thousand military personnel to the area. The troops were very much to be seen in Leighton Buzzard mainly for shopping and nightlife. I remember in particular the occasional sports encounter with members of the RAF from Wing who played the boys of The Cedars Grammar School at cricket and rugby and whom they quite mercilessly crushed. One rugby score was 89-nil, but the RAF did have a player named Mountford who was a professional in Rugby League. One other military to be sometimes seen in the town were Czechoslovak soldiers who were part of the establishment of the Czech president (Beneš I believe) who spent the war years in exile at Aston Abbotts.

One aspect of those years which brought home to us the horror of war was when people we knew or knew of were killed or taken prisoner. One particular pal of mine did not see his father, Stanley Robinson, for five years after he was captured in France with the British Expeditionary Force and remained a prisoner for the duration. Sadder still was when boys a few years older than us were reported killed. Two boys from The Cedars School – Peter Pidgley and John Hoddinott and one boy – Ken Sayell a fellow member of St Barnabas Church Choir are names that come to mind of young men who died in battle. On a few occasions there were reasons for cheering when our headmaster announced that an old boy had been decorated for bravery or had been engaged in a successful war operation. I recall the announcement of an old boy surnamed Craig being awarded the DFC and of another old boy – a sailor whose name I cannot recall being involved in the sinking of the German battle cruiser *Scharnhorst*.

On an altogether different level there was a spirit of self-help that was engendered by the government and known as 'the war effort'. It encouraged people to take part in all manner of activities. I personally collected acorns for pig food; rosehips from the hedgerows to make a substitute for orange juice for babies; and as a boy scout I collected old newspapers, which were deposited in stables in Mentmore Road. At The Cedars School, then in the centre of Leighton Buzzard, the boys were set to digging an area of ground to the south of the main buildings. These were to be the new kitchen gardens for growing vegetables. I learnt here what 'double digging' was – because of the rough nature of the ground it had to be dug two spits deep – it was backbreaking. In addition the boys at school helped with work in the school grounds, sweeping up and barrowing all the autumn leaves – under the watchful eye of the school caretaker – a 'tyrant' known as Bill Varley.

With the absence of so many men at war, children were encouraged to help where they could. I was released from school one autumn to go potato picking. This was at Grovebury Farm, run by the Heley brothers. The potatoes were turned out of the ground by a horse-drawn digger – a rotating wheel with tines that churned over the top nine inches or so of the soil. We lifted the potatoes and put them in wicker skips for about 30 pence a day. It was backbreaking work. Sadly the work came unexpectedly to an end after about five days when

Mr Poole the digger driver caught his arm in the rotating wheel of the machine; the poor man screamed for help and was in dreadful pain having broken his arm very badly.

I also did work for Mr Abrahams in a field by the canal near Three Locks. It was the very boring job of cutting down thistles in a cornfield using a wooden pole with a sharpened hook at the end. I also did harvesting on a farm during two summer holidays. This was even harder work. The first job was the making of sheaves of corn following the binder around the fields; then the pitching and carting after the sheaves had dried and finally the stacking. We used to start at 8.00 a.m. and work till dusk – about 10 hours after breaks and for a 15 year old it was extremely hard work.

Another war effort event was the raising of money, usually to buy a Spitfire aircraft. These were known as Spitfire Weeks. The whole town would hold events – raffles, dances, and so on. There would be a 'thermometer' in Church Square showing the sum raised as it climbed the scale to reach £5,000. I recall Bernard Miles the well-known actor, who was living at that time in Ivinghoe Aston coming to open proceedings in Church Square – I can see him now in a purple jacket calling on the audience to do their bit. I was in the fourth form at The Cedars School on one such fund-raising week and as school children we decided to put on two plays by WW Jacobs – 'The Monkey's Paw' and 'The Dear Departed' and charge four pennies admission to the performance. I think we raised a few pounds towards the Spitfire Fund, but mainly we enjoyed ourselves in what at that time was a noble cause.

Another memory of children helping with war work at The Cedars School was when pupils were engaged for two or three periods a week packing Sten gun components – everything from the barrel down to small screws, pins and nuts. Each component was packed in set numbers of 10, 20, etc., into cardboard boxes of varying sizes depending on the contents … The next job was packaging all the smaller boxes into large packing cases for transport. All this used to be done in the 'Conservatory' of the school. Our parents were not so pleased however having to remove and clean all the varnish splashes from our clothes.

One very interesting variation to woodwork classes at school was the making of wooden models of ships and aircraft out of normal softwood and balsa wood. The government issued scale drawings of all types of merchant ships and war ships, British and foreign and similarly aircraft. It was our job to make models strictly to scale with all details correct, all markings on aircraft etc. The models were all painted and correctly coloured even down to the decking on ships and markings on aircraft. These models – the ships up to 15 inches long and aircraft about 6 inches long were sent to RAF establishments where they were used for recognition training by airmen. At one time a big display of many of these models was put on in what was then the Gas Company showroom, now where the Sue Ryder shop stands.

The real war hardly touched Linslade, happily we were away from the bombing and so on, but we still had to observe the blackout. There were no streetlights, so the nights were very black and no lighted windows to lift the gloom. Air raid wardens would patrol the streets at night and knock on doors if a chink of light was showing. One occasionally read in the local paper of people being fined for breaking the law and showing a light through their windows. One result of the blackout – notwithstanding there being permanent daylight saving – was that church evening services could not be held after dark as it was not possible to black out the windows. Therefore they were held at 3 p.m. from about November to April.

Linslade and Leighton were mercifully spared bombing apart from just a couple of occasions. The first and only bomb in Linslade fell in late 1940 in a ploughed field just west of the railway line roughly opposite where Mentmore Road allotments now exist. Many people walked through the muddy field to view the crater about 15 feet in diameter and 8 feet deep. The only other bomb I recall was dropped at about 9.00 p.m. one night with an enormous bang. It landed right next to the old Leighton Buzzard Workhouse (now Ridgeway Court) in Grovebury Road. It made a large crater in the very sandy soil but seemed to do very little damage to the buildings. That same night many incendiary bombs were dropped over the town and countryside but caused little damage. I recall cycling out to Ledburn and looking for the burnt out remains of incendiaries that had landed in the fields round about.

Generally there was freedom of movement around the district – the only exception I recall was a very long and heavy wooden pole erected in Wing Road about 100 yards west of the entrance to Southcourt Stud. It was mounted on a large concrete base and pivot and permanently open. I imagine it was intended to be used to block the road in the event of an invasion. About 50 yards on the Linslade side of this barrier was a machine gun emplacement constructed of timber poles and sand bags – it was never used.

Everybody had to have a gas mask and in the first few years of the war you carried it wherever you went – this was compulsory. In Linslade the headquarters of the ARP was in Brooklands House (opposite St Christopher's garage) and we had to attend there to be fitted out with our gas masks. The mask was carried in a cardboard box hung around the neck on a strong cord, however this did not last long. Enterprising manufacturers soon got round to making plastic or canvas containers to put over and protect the cardboard boxes. Practices were held to train people in the use of the masks; all I can remember about them was that they steamed up quickly and you could not see out of the visor.

The Wing Road railway bridge was a strategic point – in effect the gateway to Linslade and Leighton Buzzard. The bridge carried the main railway line from London to Birmingham and the road to Aylesbury and the west passed underneath. A barrier could be swung across the road and defended by the machine-gun post placed by the Home Guard.

At the time of the air raids on London 1940–1941, although we were hardly affected, the sirens sounded the alarm fairly frequently. At The Cedars School we had to leave our classrooms and make our way down to the lowest floor of the school which housed the boys' cloakrooms and sit on benches until the all clear sounded. After a time it became apparent that the town was not the target of the German bombers and so the warning went largely unheeded. For shoppers in the High Street there were large public underground shelters in the old brewers' cellars that still exist under the service road to the rear of the Bossard Centre, the entrance being via a flight of steps about 100ft along the mall leading up from the High Street to Waitrose and adjacent to the side wall of Brown and Merry's office.

After about 1942 it became more noticeable that RAF and US aircraft were in the skies most evenings of the week flying east and north-east over the town. We knew from the news that these were on their way to take part in bombing raids over Germany. On occasions their numbers were in hundreds and I suppose we felt proud that we were fighting back in the cause of freedom. As youngsters we did not give a lot of thought to the fact that many of those planes that we watched going out would not return and many airmen would lose their lives.

On a few occasions aircraft crashed locally and we would cycle out to view them. I recall a Liberator bomber, which crashed in a field between the canal bridge and Ivinghoe, it was on an outward journey from Cheddington airfield and sadly some of its crew were killed.

Another Liberator on a homeward run from a raid over Europe crashed in woods near Edlesborough on the road to Ashridge. On this occasion the crew had bailed out. A third incident was of a Flying Fortress which crashed in a small spinney adjacent to the road from Wing to Stewkley landing upright in the trees resting on its nose. Finally, a Spitfire crash-landed in a field at Grove between the church and Grove Farm and close to the road – I believe the pilot walked away from this crash.

One other form of war service was that of the 'Bevin Boys'. I recall two boys from Linslade being directed into the mines; one was Gerald Samuel, son of the headmaster of Linslade Boys' School and the other a boy named Clark from New Road who sadly was killed in a mine accident leaving a solitary widowed mother.

The Cedars School staff was not unaffected by the war. It lost two of its younger men teachers Mr Adams and Mr Steele to military service. Older men teachers were not affected. It meant that we had a preponderance of lady teachers. One result of this was that boys' gym lessons were taken by physical instructors from RAF Leighton Buzzard who were very strict and tough on the boys – treating them rather like military recruits.

In and around Heath and Reach areas of woodland were closed to the public and used as an army training ground and in which live munitions were used. Some boys from the village did as many boys do and trespassed into the prohibited areas. They found a live explosive device and it exploded killing one boy – the village policeman's son – and injuring two others. One of these was a fellow pupil at The Cedars School by the name of Hawksford. He was blinded by the blast and I recall his badly scarred face and having to be led around as he never regained his sight. He later went to a school for the blind in Worcester where I believe he managed to continue his schooling and achieve some success.

After the tide had turned in the Allies' favour and it became apparent that they would one day be launching an invasion of Europe, one began to notice large-scale movements of troops and equipment. I recall on three or four occasions when thousands of soldiers would travel on troop carriers through Linslade and along Wing Road where I lived – on through Leighton Buzzard and eastwards to future points of departure. The convoys would last up to three or four days more or less non-stop, apart from breaks when all the vehicles rested for about 30 minutes, during which time householders would come out and provide hot water for the soldiers to make tea parked by the kerbside. The convoys included large artillery, tanks on transporters, Bren gun carriers, heavy earth moving equipment such as bulldozers (on low loaders), scrapers, dumpers, petrol tankers and so on. The surface of the road at the junction of Wing Road and Old Road outside the Bedford Arms became quite broken up as the tracked vehicles turned right around the old First World War memorial which stood there at that time.

Eventually when the invasion of Europe began the sky over Linslade was crossed by hundreds of gliders – Hengists and Horsas – towed by Lancasters, Halifaxes and Stirlings. The gliders were identified by two broad white bands around the fuselage. To us young teenagers it was an exciting time as the war was now going in our favour and there was much to enthuse about.

At the end of the war in Europe plans were made for a service of thanksgiving and remembrance. I recall few items – mainly the music – that we rehearsed at The Cedars School for the occasion. We sang 'Non Nobis Domine', 'Jerusalem', 'Land of Hope and Glory' and 'Let us now praise famous men and our fathers that begat us'. I have recollections of joining up with other schools and the general public in Parsons Close in the centre of Leighton Buzzard where a service was held. Various other celebrations were held – the one I recall was a bonfire with fireworks on Southcott village green opposite The Hare Inn. It was a joyful occasion.

An epilogue: The hard winter of 1940 brought a lot of snow and many children enjoyed sledging on Tunnel Hill, which was the best spot in Linslade. And I recall two boys three or four years older than I was who displayed flair at their speed and daring flying down the hill. They were Tom Gibbins and Toby Jordan. I can see Tom flying down the hill and crashing into the hawthorn hedge at the bottom. He would get up, his face scratched and streaked with blood and go back up the hill and do it again and again – I thought what a great fellow he was. Three or four years later Tom was reported killed in North Africa. It brought home to me the harsh and dark reality of war, thinking of a daring sixteen-year-old boy on a sledge to a young man dying a few years later on the battle field. Seventy years later I often see his youngest brother and say hello and still remember Tom.

Hans-Wolfgang Danziger

My sister Marion Goldwater and I came to England in March 1939 aged five and eight respectively. We were met at Liverpool Street Station by Miss Turner, the assistant to Mrs Sabakin, Matron of the home established by the late Lord John Sainsbury in Putney for 21 children. I remember our time there as being reasonably happy; we were well looked after but the learning of English appears to me now as a battle between the poor man entrusted to teach and his unwilling pupils.

After about three months I had to go to the local junior school. With a minimal grasp of the language I was horrified to be surrounded by poverty-stricken boys whose bottoms were often exposed, due to the torn state of their trousers and who constantly demanded piggybacks. I also remember to this day the smell of the unwashed clothes. The teacher who ruled 52 of us was Mr Fraser, a small man with a Prussian haircut who brandished a short cane as a constant threat. He told us that by the next term starting September he expected perfect English, or else … I was saved by the outbreak of war.

In August we were all taken to Bexhill for a holiday and I was able to send photographs of that time to my parents in Berlin. On our return to London at the end of the month, most children were evacuated to Reading, but my sister and I together with Marion Myrantz were sent to old Mr Sainsbury's riding stables in Leighton Buzzard, to live with the rather Victorian groom and his wife. Although I enjoyed riding the enormous hunters around the yard, Marion and I remember well the misery of trudging around the recreation ground in freezing weather to get fresh air. The smell of the iron swings is still with me and I avoided them when later looking after my grandchildren.

Luckily our teachers at the school were kind; however after a short time we were sent to Harpenden to live with the daughter of the groom. This woman of 26, and whose husband was called up, received 15s for each of us from Mr Sainsbury (this was before he was awarded a life peerage) for our keep.

Extras were always granted, like the football boots desperately needed for one game after which I was chucked out of the team. We were happy with her and glad that we could attend her funeral when she died aged ninety. Mr Sainsbury sometimes came down to see us and take us to tea. He paid the fees for my grammar school to which I moved and likewise he did for my sister's college which she attended after leaving Stoatley Rough School.

There was also the annual visit to Sainsbury's HQ for tea and to present our reports. When my parents, who had miraculously survived the war in Berlin, finally came to England, Mr Sainsbury together with Woburn House paid for their living expenses. We always exchanged Christmas cards and there was a final twenty-first birthday present, Roth's *History of the Jewish People*.

As I was unwell at the time of the death of this kind, generous and dignified man, I was pleased that at least my sister was able to attend his memorial service and speak with his family.

Peter Laufer (now Langford) has confirmed that the children who lived at 'The Laundry', Southcourt Stud, in 1939–40 were: Ernst Caspari, Bernhard Goldmann, Franz Oppenheimer, Bruno Berger, Wolfgang Zernik, Hannah Hirsch and Marion Frischauer. Stoatley Rough School was founded by Dr Hilde Lion in 1934, largely through the efforts of Bertha Bracey, a Quaker eager to alleviate the plight of German refugees. It was a mixed boarding school and catered mainly for refugee Jewish children.

Tony Pantling

Born in 1927, Tony was a pupil at Cedars School in Church Square when war broke out. He lived in a council house in Soulbury Road, Linslade. In one sense his life went on as normal: he went to school, sang in St Barnabas church choir, and took dancing lessons in a successful effort to meet girls.

His mother, like many families, took in evacuees from London and later RAF personnel as lodgers. In a small council house, with a younger sister and brother, and only an outside toilet, this was 'quite a squash'. His father had multiple sclerosis, in those days called disseminated sclerosis, and spent the entire war in hospital in Aylesbury. Tony remembers:

> Every Wednesday afternoon, until he died of influenza in February 1945, mother would catch the bus to Aylesbury to visit dad in the infirmary. Each Tuesday evening before the visit we would spend half an hour with a Rizla cigarette machine rolling ½ ounce of tobacco into about 40 cigs, which would last until the next visit.
>
> Father was unable to roll for himself as he had no sense of feeling in his hands but he could hold the cigarette as long as he could see his hand. He could not see well enough to read. His family went to see him at the weekend and I would go occasionally by bike, or by bus, or sometimes I had a lift with an agriculture rep who had a petrol allowance.

Below are other memories of those years split into themes to provide an insight into life in the two towns as the war progressed and Tony grew up. He recalls the day war broke out, the evacuees in his home, how he followed the war, bombs and other events. He started work at Barclays Bank; then he lied about his age in a bid to join the RAF but ended up in the army. Wages were low and times hard. At the right season he rose at 4 a.m. to pick mushrooms 'before anyone else got up' and sold them at 6 pennies (2.5p) a plate to regular customers. But as he points out, that added up to quite an income:

> Very best seats in the cinema were 1s 10d. Beer was about 6d a pint, Players cigarettes 1s for twenty ('Woodbines' were cheaper). A ¼lb slab of Cadbury chocolate was 4d. Our house rent was 8s 10d per week, including rates. A little cash could go a long way, but life was comparatively simple and basic, and expectations were modest. Euston was 4s 10d return third class; there was no second class in those days.

WAR BEGINS

Preparation for the conflict had been made for at least a year before. Air raid sirens had been installed (in Linslade at the police station in Wing Road), gas masks had been issued and air raid shelters built … where necessary, the civilian populations had purchased and prepared curtain materials to ensure the 'blackout' of their homes. In the run up to September 3 many reservists had been called to the colours and in a sense the whole country was prepared, even expecting, the sombre words of Neville Chamberlain when he spoke to the nation at 11 a.m. on that Sunday morning. We hoped for the best but prepared for the worst.

There was perhaps a feeling of relief that the die had been cast and now we could get on and deal with the Germans – but this was not without a certain apprehension, since the Great War had ended only 21 years before. The majority of adults had a clear understanding of the carnage and casualties of those days.

The day the war broke out is surely remembered by all of school age or above. I was in the choir of St Barnabas – Matins at 10.30 a.m. followed by the Communion Service. I clearly recall the verger, Mr Frederick Vickers, coming into the church discreetly from the door next to the organ and approaching the vicar, the Reverend Lionel Edward Lydekker, at the altar, for a hurried whispered conversation. I guessed correctly that war had been declared.

A friend of mine, Ken Sayell, who was sitting next to me, was killed in Normandy five years later when the tank of which he was a crew member was utterly destroyed. Stanley Abbot, server, and Frederick Jeffs, bellringer, also died in the war.

As I walked home from church after the service the air raid sirens alerted the town to possible air attack, and the wardens and the police officer in Soulbury Road, Mr Walkden and PC Frankham, were busy shepherding the few pedestrians into their homes and 'out of danger'. The all clear fortunately sounded within the hour, and normal life was resumed by all. The alert had been triggered by a solitary friendly aircraft approaching across the North Sea causing a possibly over-zealous 'spotter' to raise the alarm across southern England. It was probably six months before the sirens were exercised again. I can recall when the alarm was sounded in May or June 1940 – my mother not being at home – getting my younger sister and brother to sit in the pantry under the stairs in case we were bombed.

THE EVACUEES

For the next six months or so after Poland was overrun there was very little warlike activity but we had the first of a string of evacuees. Mother certainly never took in evacuees, or RAF personnel, to make money, but rather because she was very hospitable and recognised their need … There was never any shortage of people wishing to come, and mother was a smashing cook, perhaps that was why.

Our first visitors were two girl evacuees, Doris and Betty Miles, relatives of an elderly couple who were living the other side of Soulbury Road, but they stayed only a month before returning to their home in London.

Then the boys from the London William Ellis School arrived, as one unit, from their billets in Rutland … The original idea was that they should share our Cedars School premises, with the Cedars children and staff going to school in the mornings and William Ellis boys in the afternoon. This was not an acceptable idea to the Cedars senior staff, especially the Headmaster, Mr Fairbrother, and after a brief time the William Ellis School was accommodated in other buildings in the town, such as chapels, and so left The Cedars school buildings. I have heard that they were permitted to use our chemistry and biology labs, but cannot recall this myself.

Our first William Ellis evacuee was Laurence Hogstead, a six former [sic] who stayed only until he had taken his upper school certificate examination. He was a little superior, had binoculars, went riding and had very carefully polished riding boots. He was probably not used to the cramped conditions and the outside loo of a council house. As soon as Lawrence left Roy Shaw asked to come, because he was unhappy in his billet a few doors away. Roy was pleasant, affable, also from a good background – his Dad had been the first British sales manager for 'Hoover', but had died years before. We kept in touch with Roy, who became a councillor and eventually Mayor of Camden in London. Roy stayed for a year and took his higher school certificate in turn and briefly joined the Home Guard, later serving in Europe in the Royal Tank Corps and after the war spending some time with the Control Commission in Germany. In 1940 Roy would occasionally bike home to Highgate in London and back at weekends, and during that summer worked at John Pratt's farm at Edlesborough.

We had a succession of William Ellis boys, most of whom stayed for about a year, and at one time had two, Gordon Cornwell and Harry Parsons, which in our small house was a bit of a squash. William Ellis boys did not have homework but I had two hours a night – in theory – and I am afraid it was very rarely 'well done'.

One day I responded to a knock on the door to find a boy of about nine outside asking, 'Does Harry Parsons live here?' It was Harry's brother Terry who had just been billeted with Mrs Watson about eight houses down Soulbury Road. Terry eventually became better known as Matt Monro, the 'man with the golden voice' who achieved international fame in the 1960s and 1970s.

When we no longer had William Ellis boys, we billeted WAAFs. There were in succession four WAAFs: Gladys, Margaret, Joan and Chris. Gladys was an ambulance driver, Margaret and Joan were teleprinter or telephone operators, and Chris was employed on domestic duties at what had been the Albion Hotel, (at the bottom of the High Street; the back entrance to the hotel was in

Friday Street). The WAAF took it over, I think for officers' quarters. Margaret married an RAF Sergeant while living with us, and had the wedding reception in a building at the rear of the Wheatsheaf in North Street. They were followed by an Aircraftman, apparently very Irish, Paddy, who since I was already in the Army, I never met.

SCHOOLDAYS AND BOMBS

To a large extent life at The Cedars School in Church Square carried on as usual, some new faces appeared, evacuees from Grammar Schools elsewhere, but in quite small numbers and easily assimilated into normal life. I can recall only eight men on the teaching staff and none of them appear eligible for military service. An Air Training Corps (ATC) squadron was formed (1003 squadron) quite early in the war, with two or three of the masters as officers. Although I was intensely interested in the Royal Air Force, for some reason I was never tempted to join the ATC – but that left me free to join the Fire Service.

At school there were designated areas for shelter in the event of an air raid – the safest parts of existing buildings. I clearly remember when sitting in the corridor leading to the boys' cloakroom under the 'new building' which had classrooms and the chemistry laboratory, hearing the distant crump of exploding bombs on the day the *Luftwaffe* attacked Vauxhall at Luton. Daylight warnings and air raids were really quite rare and I cannot recall another while I was at school. Night time, of course, was different, and I can remember at least three occasions when bombs fell in the vicinity, and one in particular, a mixture of high-explosives and incendiaries in the Grovebury Road area, some in the Wing-Mentmore Road, and a few at the top of Soulbury Road opposite the water pumping station. For relatively small bombs, they made a considerable whistle on the way down.

The craters were only about four feet across each from bombs that were probably 100 pounds. Of course, we boys were off on our bicycles to inspect the craters and see if any shrapnel could be recovered, although I never had any.

From the top of Soulbury Road it was possible to see the glow of the fires in London from the bombing and the pinpoints of light from the anti-aircraft fire. On another night we heard formations of German bombers on their way to Coventry. On the way back they were scattered and more dispersed.

COTY PARACHUTES

Another harmless activity for boys was trying to recover the parachutes or flares made for two-inch mortars. Coty had established a factory near Southcott Stud to manufacture magnesium flares used by the army in two-inch mortars. They used to test fire a proportion of their production, possibly once a week. The flares would leave the casing at probably 300 feet, and, especially if there was a wind, the parachute would float quite a distance, long after the flare had burnt out. Local boys would often pursue the parachutes until they descended, perhaps as far away as the Firs at Stockgrove, to salvage the chute itself.

The parachutes were about three feet in diameter and made of cotton (not silk) but I suppose still quite useful for making simple underwear or similar. Clothes of course came only 'on coupons': that is they were rationed. I was never fortunate enough to find one, although the factory was not bothered to recover them, probably not worth the trouble. Coty was a cosmetic manufacturer and had the expertise and machinery to grind magnesium to the fine powder required for flares.

BATTLE OF BRITAIN AND THE FEAR OF INVASION

After the defeat in and of France the tempo of the war changed dramatically. The Battle of Britain was followed by the Blitz. The Battle of Britain had very little effect on civilian life in Leighton Buzzard. It was fought over Surrey, Sussex and Kent. We listened to the radio news bulletins and avidly read the newspapers, the *Express* in my case, for news of our successes. I recall that, probably on the decisive day of the battle, the news was that we had downed 182 aircraft, or so it was claimed. The *Daily Express* next day had row after row of tiny black aircraft above its normal headlines on the front page to represent the 182 planes shot down.

After Dunkirk the Germans began preparations to invade us and moved troops into the Channel area and began to mass invasion barges and other equipment in the Channel ports. As a precaution all the signposts were removed from our roads, and all the station names were removed at railway stations in order to make it difficult for any invading force or parachutists to know where they were, or indicate if they had reached their objectives. Train carriage lights were very subdued, and of course blinds drawn at night, so if travelling by train in a strange area it was difficult to identify one's destination, especially at night. Trains did not have intercom as today so as a train stopped at a station a porter, or possibly the station master would shout out the station name loudly so passengers knew when to alight.

Although service units would have maps supplied for troops and vehicle journeys by road detail would obviously be limited and a cross-country journey could be quite difficult to navigate successfully in the absence of all road signs. This hardly affected the civilian population whose travel was almost entirely limited to public transport because at least 90% did not own a car, although some owned motorbikes, and there was no petrol for non-essential purposes.

Official vehicles had petrol, ambulances, fire engines, police, post office etc., but civilians had to prove a need, doctors, some vets and other assorted occupations requiring vehicle travel. Most cars were 'mothballed' for the duration.

Farmers of course needed petrol for tractors along with essential delivery services that also had to operate. Horses were still in general use on many farms and for tradesmen's delivery vehicles.

THE CHOIR

Bearing in mind that I was 12 when war broke out there was little social activity I could get up to apart from cinema visits, although we were not allowed to go to the 'pictures' in term time. So mostly I was involved in the St Barnabas choir. There were three practices a week, Monday and Tuesday at 6.30 p.m., boys only and Friday 8 p.m., boys and men, a full rehearsal for Sunday. I was only a run of the mill singer but really enjoyed being in the choir. We always had 16 boys, with probationers waiting for a place as the elder boys' voices broke. There were two services on a Sunday – and three on the third Sunday. Before the war we had annual choir outings – Windsor, Brighton, Southend – I remember – but of course this finished in 1939.

We did get paid. Provided no practices or services were missed, one had three shillings a quarter. Missing one practice or service in three months meant two shillings and six pence was the maximum earned. Two senior boys were the book boys, arrived early and set out all the appropriate books on each choir stall, Psalter, hymn book and any anthem ready for the choir members when they entered their seats. I was a book boy for the last two years and therefore received an extra 1s 6d a quarter, total 4s 6d. All choirboys who attended morning service on Christmas Day received a gift of half a crown from a Mr Blows who lived near the church. It was the duty of the two senior boys to call on Mr Blows after the morning service and to thank him for his kind gift, which I did with some trepidation for he was a very austere gentleman and clearly very rich. I left the choir in 1941 and after a year or two became a 'server' until I joined up.

THE YOUTH CLUB

We had a church youth club, which met once a week in the nearby school hall. It was very simple; soft drinks, chat, gramophone records, dancing. All very innocent. After I began work in Barclay's Bank I became treasurer of the club, monthly collecting and accounting for admission payments and drinks etc. Very simple. I also joined the youth club at the Hockliffe Street Baptist Chapel, which had an excellent cavernous meeting room below and was altogether more suitable, very well attended and organized and of course mostly boys and girls from Leighton Buzzard instead of Linslade. Some of the boys I knew from school and some like Don Willis I know to this day.

THE YOUNG COMMUNIST LEAGUE

We also had a Young Communist League in Leighton Buzzard, which I again joined, again being treasurer – although there was not much to treasure. We met perhaps fortnightly in an upstairs room behind the Black Horse in North Street. Such funds as we had between us we used to send writing paper and pens and pencils to our comrades in the Russian Army. It was a very small club. In fact I remember only two – Godfrey Turney and Gordon Woolford – Cedarians in the same year as myself. Godfrey used to daub graffiti 'Open a second front now' on various prominent walls, but I was never involved in this.

DANCING

It was a definite social asset to be able to dance – ballroom – certainly one needed to quickstep, foxtrot and waltz. Miss Peggy Cook used to hold dance

---

Announcing Two More WAR-TIME DANCES
at the EXCHANGE THEATRE, LEIGHTON BUZZARD,
on SATURDAYS, JUNE 29th and JULY 6th.
Dancing from 7.45 till 11.45 p.m.

FRED GROOM AND HIS FULL BAND.
Featuring new and popular programmes for Band and Vocalists.

Admission Two Shillings; H.M. Forces in Uniform 1/6.

EXCHANGE THEATRE, LEIGHTON BUZZARD.

GRAND DANCE
on
FRIDAY, 5th JULY, 1940, 7.45 until 11.45,
IN AID OF
RED CROSS
(Central Hospital Supply Service).

LAWRENCE INNS AND HIS BAND,
the members of which are giving their services free.

Tickets 2/6, Members of H.M. Forces 1/6,
can be obtained from Barclays Bank Ltd., Westminster Bank Ltd., Midland Bank Ltd., F. W. Woolworth & Co., Lawrence Inns, C. W. Tucker.
BALCONY RESERVED FOR DANCERS.

THURSDAY DANCES
at the
CORN EXCHANGE, LEIGHTON BUZZARD
(by permission of the Commanding Officer),
JULY 29th, 8 p.m. to 11.45 p.m.
FRED JANES AND HIS DANCE BAND.
Tickets 2/6, H.M. Forces 2/-.
From Messrs. Woolworth's, Lawrence Inns and Unicorn Hotel.
IT'S BOUND TO BE GOOD.

A wide variety of social events was available to boost the morale of civilians and service personnel alike. Dances were a popular distraction, keeping both mind and body busy; several were held each week in any available venue.

classes in a small studio in an upstairs room in Water Lane (the building is still there, on the left hand side, before Tudor Court). It was one shilling a session I think, but I also had some private lessons at additional cost. I made sufficient progress to ask any girl to dance, and the expense of these lessons paid me dividends in enjoyment for years to come. Ballroom dancing was THE way to meet the girls. You noted a girl you fancied, asked her to dance, and hey presto, you are in an intimate one to one situation in no time, and a street ahead of the fellow who cannot dance, or is inept, and so no pleasure to dance with.

There were often two or three dances in Leighton and Linslade on a Saturday night, the Old Vic, Co-op Hall, Forster Institute, all well attended because of the service men and girls in the town, also Land Army girls billeted in Hockliffe Road and Harcourt House. The RAF came not only from Stanbridge and also HQ 60 Group but also the airfield at Wing. Such army as we had in the area were insignificant by comparison. There was not a lot of drunkenness, probably because the pubs closed at 10 p.m. American servicemen were very rarely seen from their base at Cheddington. Perhaps they went to Aylesbury.

Some American officers working at Bletchley Park were living at the Hunt Hotel in Linslade. In December 1943 my mother had a part-time job cooking and waiting at tables in the Hunt and was given a Christmas gift by an American resident – a large white and black crisp £5 note. It was the first one she had ever held in her life, let alone owned. She had two weeks' income in one note.

TRAVELLING FAIRS, SERVICES CANTEEN AND BANK PRECAUTIONS

Travelling fairs still visited Leighton Buzzard in war time but due to black out requirements were probably working in the summer only. Of course in those war time days we had 'double summer time'; in other words we advanced the clocks two hours instead of one. The main intention was to assist the farming industry, but we all enjoyed broad daylight, in the height of summer until at least 11.00 p.m., and of course the Fair could operate almost normally for those months. I well recall one visiting Bridge Meadows in September 1943, the works, roundabouts, dodgems, swings, sideshows, etc. I was introduced to a girl friend on this occasion, rang her recently at her home in Goring on Thames to confirm this was so – not bad after 70 years.

There was a Services Canteen in the High Street, above what is now Costa Coffee, in Ravenstone Chambers. It was quite large because as well as serving refreshments they also had a table tennis table, darts etc. I went there once only, after I joined the Army, but of course I lived locally and did not need the facility as visiting service people did. I think it was run on a voluntary basis by the W.V.S. It would, of course, have been very inexpensive; basic pay for a private soldier, or an aircraftman in those days being three shillings per day.

We had three banks in Leighton Buzzard. Barclays, Midland (now HSBC) and Westminster (now NatWest). There was concern that should a bank be bombed, then all records of current transactions and balances could be lost. Therefore at close of business each day a summary of that day's activities and current balances (appropriately sealed or locked to maintain confidentiality) was taken to another bank in the town for safe keeping. When I worked at Barclays, I would take ours to Midland or Westminster and they would bring theirs to us. At the bank the cashiers would routinely take out of use all the 10 shilling and £1 notes that were distressed or badly damaged; these would be saved until we had amassed a large quantity, when they would be securely parcelled up, and the parcel sealed with sealing wax. I then walked down to the post office, to send it registered post to the Bank of England for the notes to be destroyed. Probably £8,000–£9,000 in my hands – 100 years' wages. There was no Securicor in those days.

FARM AND OTHER WORK

Although farm workers were in a 'reserved' occupation, and therefore not called for military service, there was a shortage of labour on the farms, and especially at harvest time. One year volunteers were needed to harvest the potato crop at the farm near Brickhill and about eight Cedars boys, myself included, volunteered to help and were given time off school to go. This was back breaking work and not enjoyable. I would rather have been at school. However, in the summer holidays of 1941 and 1942 I spent many happy weeks working at John Pratt's farm at Edlesborough known as Sparrowhall Farm, biking from home to start at 8 a.m. I was paid 6d (old pence) an hour, the government rate, 25s a week for 50 hours of getting the harvest home.

At Easter 1942 I was a temporary telegram boy based at Leighton Buzzard Post Office in Church Square, delivering telegrams by bicycle. The furthest I went was Stockgrove Park to some sort of secret military establishment and sometimes to 60 Group, RAF, in Plantation Road. It was a simple task, paying about the same as farming. In December 1942 I delivered the mail for about two weeks in South Street and Stanbridge Road areas. Two deliveries a day but one only on the morning of Christmas Day. Pay again was about 6d an hour. I earned one shilling in tips and two people gave me 6d each.

GIRLS' SCHOOLS

Prior to the war a private school for girls (and I think boys up to seven or eight) had been established at 'Woodlands', a large country house with considerable grounds in Soulbury Road, bounded by the bridle path and Linslade Wood. After the war started a school from London joined them for a while, and then moved as a separate school to the Gables in Linslade. This was the King's

House School from Highgate. When this school eventually returned to London another private school was established in the Gables for some years after the war.

Meanwhile some of the King's House schoolgirls appeared at morning service at St Barnabas and were of considerable interest to the choirboys, or at least this one, although they were so carefully chaperoned, and never left the Gables alone. It was very difficult to communicate with them. Nevertheless I knew at least one, Winifred de Silva, quite well and established a link that lasted some years, and I eventually stayed with her family in London. The Woodlands School lasted for a year or two but eventually closed as the buildings were used for prisoners of war.

RAF VOLUNTEER RESERVE

I was very keen to get into the Royal Air Force, and in 1943 'lost' my identity card. Everybody had to have one of these and carry it at all times. I applied for a new one at the Food Office in Bridge Street that was normally involved with ration cards. In seeking a new card I said I was born in 1926, and was therefore already 17 and a half. My new card, showing my new age was issued accordingly and I could now apply to the RAF to volunteer for aircrew training.

I was sent to Mill Hill for a very searching medical, and then to Cardington for a three-day selection course. I was not a member of the Air Training Corps, most of my fellow applicants were NCOs in the ATC but fortunately I had a very great interest in aircraft British, American and German and a basic knowledge of navigation. As a result of this knowledge I was one of four out of 40 applicants to be selected for training as pilot navigator and bomb aimer. About another 10 were selected for wireless operators and air gunners, known as WOP/AG's [sic].

However, I had a problem. I could not be sworn in and become a member of the RAF volunteer reserve because two doctors said I must have my tonsils out, despite my protest that I had them removed in 1931 at the Royal Bucks in Aylesbury.

I returned to Leighton Buzzard, saw my doctor, Dr James, and told him of my problems. He got me into the Royal Bucks again, I had my tonsils removed again and was back at RAF Cardington within a month, and was duly sworn in and given an RAF number 1884002 – AC2 (Aircraftman second class) Pantling was me. I was told to go home, wear my RAFVR (Royal Air Force Volunteer Reserve) lapel badge and await call up.

The manager at the bank had been very understanding about all the time off for interviews, hospital etc. This would be in April 1944 (before D-Day in June). In September that year the RAF called me to another selection board at the Grand Hotel in Torquay, the building an empty shell apart from the basics of an RAF unit, bunk beds and the necessities of service life. Similar tests to those at Cardington, possibly more stringent followed. Two weeks later I received a letter from the Air Ministry saying they regretted my services were no longer required, and soon I received my call up papers for the army. Hence at the age of 17 years and five months I became a soldier at Northampton. The entire company that I joined had been to Torquay for selection board recall – 180 of us. It appeared we had all been rejected: the selection board had merely been [going] through the motions.

12

KEEPING THE TOWN SAFE

Local Defence Volunteers and the Home Guard

In the late spring of 1940 it was widely thought that German paratroopers might be dropped on Britain, possibly disguised as policemen or nuns. Q Central was defended by the RAF regiment and there were a number of army units in the area, but there was no defence force for the civilian population. On 14 May Foreign Secretary Anthony Eden appealed on a BBC broadcast for volunteers to join the Local Defence Volunteers (LDV) to resist any such infiltration.

After hearing Eden's appeal, Hugh Aitken signed on at the police station in Leighton Buzzard and swiftly became a member of the Billington LDV (later the Home Guard):

> Our main duty in those first days was to watch the surrounding countryside from our dug-out at the top of Billington hill in the village. There was a nightly patrol to guard against an enemy parachute drop in the vicinity. The RAF Signals Station, Q Central, was only a mile or two away from our position, and Bletchley Park Code and Cypher HQ was only seven miles distant. We patrolled for two hours followed by four hours off, and our sleeping quarters were an old gypsy-style caravan which was quite bug-ridden and needed to be fumigated from time to time. As it was, we frequently went off duty to do our daily work madly scratching ourselves. All were issued with (at first) an LDV armband and a wooden pole (about six feet in length) very similar to the 1930s Boy Scouts' staff, and that was it.

Another platoon was also formed at Eggington. The men were issued with uniforms and met several times a week at the Red Lion in Hockliffe. They were trained to use tommy guns and sten guns at a firing range in Luton. They also learned how to throw hand grenades. Rolls of barbed wire were erected round hedges in the village and iron spikes placed on Charity Hill in the village to stop any enemy tanks that might come that way. Home Guard HQ consisted of a mess and Nissen huts; and two concrete bases for gun turrets can still be seen among the hedges there.

A public notice announced that the Leighton Buzzard branch of the British Legion had decided to form an LDV detachment and a meeting was held at the British Legion Club in Market Square on 30 July 1940 to discuss arrangements. Members and all ex-Service men were invited to attend. The first 'officer commanding' of the Leighton Buzzard Home Guard was Colonel James Macarthur, a veterinary surgeon who had served in the First World War and lived at Stafford House in Leighton Buzzard. He died in 1941. One recruit, Viv Willis, recalls that when he joined the Leighton Buzzard LDV there was one rifle for thirty men. As part of the training, each volunteer was allowed to fire the gun once. Viv was one of them: he said it was the only time he ever fired a gun in his life.

Macarthur was succeeded by Dr Thomas Crockett, the headmaster of the William Ellis School, who was appointed major. Initially the Leighton Buzzard and Linslade Home Guard platoons operated as separate units but in May 1943, by which time they had had a proper issue of kit, they were amalgamated into 'E' Company of the 6th Battalion of the Bedfordshire Home Guard under Major Leslie Francis Vick, MC, who had taken command on the death of Major Crockett. The 6th Battalion commander was Lieutenant-Colonel Dealtry C. Part, the Lord Lieutenant of Bedfordshire, succeeded later in the year by Major William Henry Cadman of Dunstable.

As in *Dad's Army*, the television comedy about the Home Guard, Major Vick was manager of a bank, in this case Barclay's in Leighton Buzzard High Street. He left the town in January 1943 on appointment to a similar post in St Albans and was presented with a framed photograph by the officers of his company. In the speeches during the presentation, Lieutenant W.S. Higgs said he had not always seen eye-to-eye with Major Vick on Home Guard matters; and Lieutenant W.R. Cox said that whatever Major Vick 'had undertaken had always produced something just above the average' according to a report in the *Leighton Buzzard Observer*. One feels that too would have fit into the script of the comedy series and it might have been better received if he had left out the 'just'.

Later Captain C.R. Kingston took over as commanding officer, with Lieutenant W.R. Cox, QC, as second-in-command. On promotion to major, Cox succeeded Kingston and was himself suceeded by Major Frank Croxford, MM. Parade Hall in West Street (now Bossard Hall) was used extensively by 'E' Company, consisting now of four platoons, nos 23, 24, 38 and 39, until the end of the war.

There were parades twice a week and absence from Home Guard duty was evidently taken very seriously. Despite an appeal for common sense from Major-General Sir (Frederick) George Wrisberg of the War Office, one private, Rupert

The Home Guard had to dig all their own defensive positions. Trenches were dug on Billington Hill outside the town and the positions for the Spigot mortar at Linslade Wood.

The 3rd Battalion, Company B, Home Guard parading down Leighton Buzzard High Street in 1942. Commanding officer Captain Stone is followed by second-in-command Oliver G. Pike.

John Groom of Rock Lane, Linslade, was fined £5, and £1 17s costs, for not turning up for a parade as his employer asked him to carry on with urgent production work. Groom's employer was also fined £2 by the magistrate.

But there was fun too. On Friday 24 January 1941, Wesley Lee and his New St Louis Players supplied the music at the Corn Exchange for a Home Guard dance, a welcome diversion from the normal routine. Most of the platoon officers were there; the platoon commander, Major Crockett acted as MC and appealed for more volunteers to join the Home Guard. Edgar Hart was responsible for the decorations and guests included the Chairman of Linslade Urban District Council, L.B. Faulkner.

When the Home Guard were stood down in December 1944, the men received no gratuities or medals but they were allowed to keep their battledress and their boots. 'Buzzwords' in the *Leighton Buzzard Observer* of 8 June 2004 expressed the opinion that the Leighton Buzzard Home Guard had achieved nothing of importance and suggested that everything would have been the same without them. More positively, King George VI, in his 'Stand Down' broadcast, referred to the Home Guard as an example of how free men would always answer the summons when freedom was threatened.

The Air Raid Wardens

In 1938 the threat of possible enemy air raids led to the establishment of the Air Raid Precaution (ARP) service. County districts were asked to form local groups including wardens, first-aiders and ambulance drivers, and to provide air raid shelters and organise storage of supplies. On the outbreak of the war the government decided that there should be some sort of defence system in every town and village in case of air attack. Many businesses in Leighton Buzzard formed night-watch groups, working on a rota system. An air raid siren was set up at the police station.

A typical ARP unit was set up in Eggington; the first Air Raid warden's post was set up at Eggington House. It was later moved to Plough Cottage, home of ARP Warden Frank Osborn. Gaius Sear and David Piggott were the other wardens, with Charlie Day as Special Constable – he came on duty whenever an air raid siren went off. An air raid shelter was dug in the garden of No. 11 Council Houses, which was a paddock at the time. Seats were put in the shelter but it could not always be used as it was liable to flooding. There was a shelter behind Patch Cottage and another at the rear of Edwards' premises. The ARP Wardens, who had no special uniform – just a badge and armband – patrolled the village at night to ensure no lights were showing through the blackout curtains.

ARP posts in the Leighton Buzzard area were eventually established at the following sites: the yard of the Unicorn (now The Lancer), 17 High Street

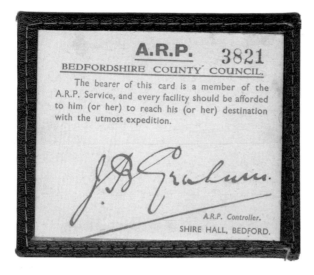

The official card carried by air-raid wardens issued at the Shire Hall, Bedford, to Miss E. Plummer of 70 Vandyke Road, Leighton Buzzard. The card is in mint condition in its own purpose-made wallet, and Miss Plummer's name is written on the back.

(formerly at Barclays Bank and the Unicorn Hotel), 'Clent' in Billington Road, 43 Hockliffe Street, 'The Chase' in Heath Road, 40 Lake Street, 91 Stanbridge Road, 42 Church Street, 121 Vandyke Road, the Mill at Beaudesert and 42 Church Street. In the surrounding district, ARP posts were established at Billington Manor, the Institute in Eaton Bray (formerly at 'Bingles'), Plough Cottage, Eggington (formerly at Eggington House), the Village Barn in Heath and Reach, the Watling Street in Hockliffe, Station Road in Stanbridge (formerly at 'The Five Bells'), the Old Smithy in Tebworth, the Social Club in Toddington High Street, the Church of St Mary the Virgin in Eaton Bray, and 'Woodlands' in Tilsworth. Pat Griffin, who was born in Linslade, recalls that her father, Hubert George Griffin (1910–86), was an ARP warden at nights and weekends. Viv Willis' father was also an ARP warden who patrolled Vandyke Road.

A civil defence HQ was set up in the former workhouse in Grovebury Road to store gas masks, housing a wartime ambulance station and providing rescue parties and a decontamination squad. First Aid Points were set up at 3 Leighton Street in Woburn, the Methodist chapel in Eaton Bray, the Schoolrooms in the Methodist chapels at Eggington and Stanbridge, the Village Barn at Heath and Reach, the Men's Club in Hockliffe, and 76 North Street in Leighton Buzzard.

In February 1940 the ARP Regional Office sanctioned the conversion of two cellars in Leighton Buzzard High Street to public air raid shelters. The two cellars chosen were in Herbert Cyril Tooley's house at 4 Market Square and Walter Pratt's premises in the lower part of the High Street (No. 15, later demolished to make way for Waterborne Walk). The two shelters were supposed to house 300 people in all. Mr Tooley's shelter was brick-arched and had no heavy overload structure to support; but Mr Pratt's cellar needed extensive strutting to make it suitable. Both shelters were equipped with electric lights, seating and lavatories. A government grant covered 65 per cent of the cost (about £300), the rest coming from local rates.

Communal shelters were also authorised in June 1940 in the Stanbridge Road neighbourhood, using plans prepared by Bertram Harold Robjant, Surveyor to the Urban District Council. The recreation ground bridge and a field path were also widened to provide an alternative route from Grove Road to the ARP HQ in the old Workhouse in Grovebury Road. The village air raid shelter at Billington was at what is now No. 1 Hillview Lane.

In September 1940, Samuel Robert Dillamore, the Sub-Controller of No. 3 Area in Buckinghamshire, through the Linslade and Wing ARP Joint Committee, advised every householder to place a bucket of sand or water – or both – outside their front door before retiring. This was to deal with incendiary bombs.

Several firms in Leighton Buzzard offered to install air raid shelters at private addresses. R. Willis & Son of The Nest, 125 Vandyke Road, offered small six-person brick-built shelters at moderate cost, 'designed from personal experience of blast effects in the last war'. Stonehenge Bricks Ltd of Mile Tree Road offered bricks 'tough as steel' for shelters. Pets were not forgotten. The RSPCA advised cat and dog owners to construct a splinter-proof shelter for their animals out of a large dustbin, half buried with soil lengthwise in the garden and heaped over with earth to a height of about 12in (30cm).

At less than £10 a week, mostly for administration, the cost of ARP services in Leighton Buzzard in February 1940 was remarkably small compared with Bedford, Luton and Dunstable. There were no paid wardens as the services managed with enthusiastic volunteers who put hours of leisure time into training for emergencies. In comparison, Bedford was spending £1,500 a week on salaries.

The Bombing Begins

Eggington, 3 miles (5km) east of Leighton Buzzard was bombed on 26 September 1940 at 3 a.m. and the windows of a house on Leighton Road (the A4012) were blown in, but there were no casualties. Subsequently two 1,000lb bombs were dropped along the same road, one of which did not go off. The road was closed for some time until the unexploded bomb was detonated. Two oil bombs were also dropped in the Pit Field behind the council houses and set a hedge on fire. Soon afterwards twenty-six small 250lb bombs were dropped on Briggington on the Hockliffe Road, a mile from Q Central.

In Tilsworth bombs fell only away from the village. Hockliffe was not so lucky as the German planes used the A5 as a guide to Coventry, and bombers flew over the village en route. On 3 October 1940 a low-flying plane passed

in mid-afternoon from north to south over Watling Street and dropped several high-explosive bombs on Hockliffe, causing considerable damage to property. Joan Holmes of Tilsworth recalls seeing the ruins at Hockliffe on a Sunday evening walk and noticed a doll sticking pathetically out of the debris. The raider then flew over Dunstable, spraying the streets with machine-gun fire.

In November 1940 an oil bomb was dropped on Grovebury Farm, Leighton Buzzard, on the 1st; high-explosive bombs on Charity Meadow and near Stanbridge Road at Billington on the 6th; an oil and two high-explosive bombs at Wood Cottage, Potsgrove, on the 13th; and more high explosives on Watling Street (A5) near the Fox and Hounds public house and Heath and Reach on the 16th.

The 'Moonlight Sonata' raid on Coventry (14–15 November 1940) was the heaviest bombing raid launched thus far. Over 400 German bombers pounded the city into rubble. Half the houses in the city were damaged or destroyed, most of the shops were ruined and casualties flooded the hospitals. A glow in the sky from the fires in Coventry could clearly be seen from Leighton Buzzard. The fatalities were officially put at 506, but the real total was probably nearer 1,000. A new word was coined: to 'coventrate', meaning to raze a city to the ground. The king visited Coventry the following day and stood in silence among the ruins of the cathedral.

On the same night, three high-explosive bombs and forty incendiaries, probably jettisoned during the Coventry raid, fell over a wide area of Leighton Buzzard, causing ten casualties of whom three were admitted to Luton and Dunstable Hospital. The bombs caused considerable damage to houses in

An incendiary bomb of the type that fell on Heath and Reach in January 1941 the town. Several small fires were started, which the emergency services soon extinguished. One 500lb high-explosive bomb narrowly missed the Grovebury Road Control Centre and First Aid Post in the old Workhouse creating a large crater only 70ft (21m) from the First Aid reception room. Although the personnel on duty were shaken, there were no casualties and because the sandy soil absorbed the blast, the building suffered only minor damage. Hazel Parrott, wife of a former POW in the Far East, recalls that people were charged tuppence (two old pennies) to look at the crater. Other bombs landed in Beaudesert, the Assembly Room in Lake Street, the corner of Dudley Street and Lake Street, the Methodist Chapel schoolrooms, the iron foundry, Grove Road, Grove Place, the recreation ground, Grovebury Road, Gossards works, the gas works, 26 East Street, 14 Lammas Walk and the Public Assistance Institute in North Street. Two locals, George Harry Chambers of Hockliffe Road and Charles Henry King of Vandyke Road, died in September 1940 when a bomb hit their place of work at Vauxhall Motors in Luton.

One daring act of bravery occurred on the night of 1/2 October 1940, at Wing, where a unit from the 27th Searchlight Regiment of the RA under Lieutenant Paul Rodney Gwyther Andrews, on his first posting, had the duty of identifying and reporting enemy aircraft sightings. A string of six bombs fell over Wing from a German plane that may have been damaged and the men of the RA unit found five craters in parkland where bombs had exploded – and one near the schoolmaster's house, where the bomb had not exploded. Although bomb disposal was not his responsibility, Lieutenant Andrews – a committed Christian with a great faith in the just cause – was very indignant that people should have to leave their homes during the night because of enemy action. He began carefully excavating the bomb, much to the consternation of the local policeman, who had to be restrained by the officer's men. After digging out the bomb, Lieutenant Andrews and a volunteer carried it to a depression where the bomb would do no damage if it exploded. Incredibly, Andrews then fetched a bicycle repair kit from his unit's HQ and used a screwdriver to remove the fuse and make the bomb safe. For this truly remarkable deed he received the George Medal. He remained in the army until 1958, when he retired with the rank of major. A man of athletic build, he gained the nickname 'galloping major' after he broke several records in Hong Kong after the war for running to the tops of local hills. True to his faith, he entered the Anglican Ministry in 1976 and was ordained as a priest two years later. Major the Rev. Paul Andrews died in 2001 at the age of 85.

On the night of 10 December 1940, the cricket ground at Soulbury became the target for a marauding German bomber when an inquisitive resident drew back his blackout curtains to obtain a clearer view. The light immediately attracted a direct hit and the house was completely destroyed; fortunately no one was killed. Another bomb, plus incendiaries, fell in a nearby field, and although subsequent reports suggested that the device had been dealt with by a Canadian bomb-disposal unit stationed in the area, there was considerable doubt at the time. After the war, the suspicions regarding the unexploded bomb at Soulbury were finally vindicated, when, in 1946, a 500lb (227kg) bomb fitted with two clockwork fuses was found near Clay Hill Farm, only 60yd (55m) away from three cottages. No. 22 Bomb Disposal Group was called in to detonate it and, the cottages being protected by a wall of sandbags, 'not even a window was broken'.

The first fatality in the Leighton Buzzard area from air raids came on 21 January 1941 when a solitary raider dropped a high-explosive bomb and several incendiaries on Heath and Reach. In an extraordinary piece of bad luck one of the incendiary bombs fell on the shoulder of a young woman, Mavis Audrey Williams, age 18, and killed her instantly. Several other people, including two evacuee children, were injured in the raid. Mavis Williams was renowned locally for her brave rescue in May 1939 of a 5-year-old boy from the Grand Union Canal. She had plunged twice into the deep water before bringing him to the bank at Old Linslade.

After 1941 there was very little damage to the town from air raids. Among the few incidents reported, several incendiary bombs were dropped in 1943, causing damage to a furniture shop known as Goodliffe's Ltd in North Street. Viv Willis recalls that when he was in Soulbury one day he saw a German plane (possibly a Dornier 77) with clear black crossings on the wings following a train along the railway line. He believes the train was hit and two people killed. After dropping several bombs near Hockliffe Road, the plane was shot down over Ware in Hertfordshire. Alan Wilsher remembers one night when bombs landed in the middle of the road just past the Memorial Hall in Totternhoe. Years later, he could still see where the road had been repaired and the telegraph wires were joined. That same night, bombs fell in Totternhoe chalk pits near the lane to Sewell. Incendiary bombs also set Jessie Bird's chicken house on fire and a bomb fell near Middle Path at Eaton Bray. Wilsher believed the damage was caused by a lone bomber who had got lost and was dumping his bombs before heading home.

The National Fire Service

During the war, as part of the ARP, a NFS was formed. It was based on and grew out of the Auxiliary Fire Service originally formed in 1838. Every village had its own unit but it did not have a good reputation in some places. The one at Tilsworth was immediately christened the Nannygoat Fire Service. The equipment consisted of a big green handcart with a pump mounted on it. One day, Celia Sinfield recalls, the thatch on Thanet Cottage caught fire and the village fire service went rushing down Dickens Lane with its cart. Unfortunately, a stone got thrown up and jammed the pump so they had to wait for the Dunstable Fire Service to come. The old joke ran 'Keep it [the fire] going till the fire service gets here.'

The NFS also had a Linslade station in Springfield Road. It was in the old stables, coach house and grooms' living quarters of Perth House, which was in Soulbury Road. This station was manned from 8 p.m. to 7 a.m., three crews taking shifts in turn. They had 3-ton trucks with room for crews and each towed a pump – a Coventry Climax – and had room for hoses, filters etc. There seems to have been little coordination between the units at Linslade and Leighton Buzzard, perhaps because the former was in Buckinghamshire and the latter in Bedfordshire, although in the daytime, when things were quiet, the Leighton Buzzard brigade was officially responsible for Linslade. In total, the two units could muster eighty men in September 1940.

Tony Pantling joined the NFS when he was still at school in late 1942 (aged 15), on duty one night in three. He was a messenger, not a fireman, but he had the same uniform, complete with equipment, axe etc. His duty was to fetch the next crew in, in the event of a crew going out. This meant going to the home of each fireman to get them to report to the station as soon as possible to man the standby pump in case of another call. None of the crew had telephones in those days. Having alerted the next crew, he had to return to the station to man the telephone.

Some of the firemen were connected to St Barnabas church because the station commander, a Mr Jordan, was a prominent church member and server. Tony Pantling's section leader was Mr Dimmock, a manager at the Coty factory in Southcott. He was paid 4s (20p) a night subsistence allowance, which he received monthly. For the last six months of his time with the fire brigade, after another messenger, Sam Sinfield, joined the armed forces, he worked two nights out of three, so he received twenty times 4s or £4 a month subsistence. He started at Barclays Bank as a junior in August 1943 and was paid only £90 a year, so the extra subsistence was very useful. All crewmembers with whom he worked were in their late thirties or older and not liable for military service.

The Auxiliary Fire Service also organised social events. On Thursday 27 February 1941 a social evening was held in the Co-operative Hall for Linslade personnel, wives and friends. The turns included Mrs Spooner (monologues), Messrs Spooner and D. Bignell and Commandant S.G. Cook (songs), auxiliaries C. Dean (impersonations) and R. Holliday (accordion), and a comic sketch from a four-man team of A.J. Aris, C. Reeve, A. Rollings and E. Baker. Games were played, and A. Rollings won at musical chairs. Fred Groom's band supplied the dancing music and Mr and Mrs A. Rollings won a spot-prize competition. Auxiliary Rollings seems to have had a great time.

Linslade's auxiliary fire brigade was on duty every night throughout the war; the firemen are seen here checking their equipment. The picture is believed to have been taken outside the council houses on Soulbury Road.

The Leighton Buzzard fire brigade outside the Old Town Hall in 1940 under Captain Baker (front row, left-hand side).

The WVS in Leighton Buzzard

The Women's Voluntary Service, better known as the WVS, was formed in 1938 to assist in the event of bombing attacks. On the day war was declared its membership stood at 165,000 nationwide. As already described, their duties included the allocation of homes for evacuees, but extended to first aid and distribution of food and clothes for people made homeless by air attacks. They also provided mobile canteens, serving refreshments to the emergency services.

The WVS helped with many activities, including book drives, organisation of the selling centres in the special salvage campaigns, assisting the food office in the issue of new ration books, distributing cod liver oil and orange juice to children and expectant mothers, surveying the air raid shelters in 1940, rosehip collections, opening a clothing exchange, and they were also responsible for the children's playground scheme and house-to-house collections.

The following extract from the *Leighton Buzzard Observer* at the end of the war explains their role further:

> The story of the WVS in Leighton Buzzard can now be told. Looking back on the six years' wartime record, one can indubitably class it as an achievement second to none. When the movement was formed in 1938 none could have visualised how wide would become its ramifications and scope, how indispensable to the war effort the work of these willing volunteers, many of them married women with home ties and other domestic responsibilities, would be.
>
> They have been aptly named 'the Jill of all trades'. Their versatility and adaptability to do the many tasks they were called upon to perform was an outstanding feat of organisation.
>
> Whether in home, communal, national or international affairs, most of us will have found from experience that when a crisis or emergency arises there is usually one person whom all depend upon to stand as a sure shield … No matter the job, the WVS were there and … could be relied upon to fulfil all requirements. At Leighton Buzzard, they were under the able leadership of Mrs Watling, ready for all emergencies.
>
> With a membership that never exceeded 200, this small army of workers put in thousands of hours, both for the Forces and civilians.
>
> They manned the canteen in Lake Street, just closed, where during the six years of war nearly a million beverages and over half a million snack meals were served. At the WVS Headquarters in Grovebury Road there were quiet rooms for the forces, and where they could also have a bath. In three and a half years the number of baths was 3,618.
>
> The members also 'mothered' the services in other directions through the work party, who undertook mending, made 200 First Aid kits for the Local

Defence Volunteers (later the Home Guard) and lots of sniper suits for the Forces. The work party also made numerous garments for evacuee mothers and children.

One of their earliest tasks was in connection with the big official evacuation of school children from London. At the outbreak of war Leighton Buzzard was a reception area for quite a large number. The WVS undertook the canvassing of billets for unaccompanied children and assisted in the reception and housing of these in August 1939. Then they were suddenly and unexpectedly faced with the problem of mothers and babies arriving as well.

The WVS met this, as they did many other emergencies, in the spirit of 'help your neighbour' and by September started a club for them to join. On Sundays they ran a centre in the Temperance Hall where evacuees could come and meet their relatives, and children their parents.

They arranged Christmas entertainment for the children and in the first year practically all the available halls in the town were commissioned for the purpose. The local children were included. The WVS also undertook the distribution of the American Red Cross parcels.

When the Council, in co-operation with the Ministry, opened a hostel for difficult boys at South View, Hockliffe Road, the help of the WVS was invaluable for furnishing the whole place and assisting in many ways. Between August and October 1940 they operated a rest centre at the Institute receiving 1,350 homeless persons from the blitzed London area and cooking over 2,500 meals during the period. Later, in June 1944 when the flying bombs came, another rest centre was opened in the North Street Methodist schoolroom.

In addition, the WVS had to cope with various problems concerned with homeless people and they were responsible for the issuing of the American Red Cross clothing.

When the Lake Street Baptist schoolroom was opened as a wartime nursery for children of parents who had to go out all day on war work, the WVS helped to get it going. They made the sunbonnets and tablecloths and assisted in the staffing.

As an information bureau, the services of the organisation were in keen demand. On this subject Mrs Watling said to one reporter: 'Everybody seemed to come to us with their troubles, it didn't matter what it was.'

One of their happiest jobs came at the end of the War in Europe, when they were called to Wing aerodrome for the reception of prisoners of war from the continent. Members of the WVS were there almost day and night, and they served the men with refreshments. In nine weeks 32,000 repatriated prisoners landed there.

This review of the Leighton Buzzard WVS activities may not be very complete in detail, but the narrative is sufficiently comprehensive to show just what a vital part this movement was called upon to play in a variety of ways to help forward the war effort.

Exercises

Before 1941 the various services held separate training sessions. On 31 March 1940, for example, according to the following Tuesday's *Leighton Buzzard Observer*, a series of street exercises was carried out over three hours to test the efficiency of all sections of Leighton Buzzard ARP personnel. In the High Street an incendiary bomb was supposed to have set fire to a hotel. The Town Fire Brigade was called out and within five minutes the engines were pumping water from the Mill Stream. First-aid parties attended the 'injured' and a decontamination squad, police and wardens dealt with a suspected 'liquid gas bomb'. After the exercise, the Chairman of Bedfordshire County Council, Sir Thomas Keens, congratulated Leighton Buzzard on 'having such a fine voluntary personnel'.

Not all the exercises were successful, notably the larger ones. Area exercises were held in the autumn and a combined exercise, nicknamed 'Scorch', 6–7 December 1941 was an example. Bedfordshire County Council's 'History of Civil Defence 1935–45' records that this was undertaken to test the anti-invasion plans of Eastern Command, particularly the information and communications systems, the defence arrangements of airfields, and the operation of the Home Guard and the Civil Defence Services. The whole county was involved. The exercise proved to be too large and too unwieldy: some areas were overloaded with incidents, while others had little or none, their volunteers standing by for long periods.

Combined exercises were held at Leighton Buzzard on 20 November 1942. These again were designed to test the military and civil defences of the area, but even at this level they proved relatively ineffective, since the tendency was for the various services involved to work separately. On this point, the rivalry between the Home Guard and the air raid wardens in *Dad's Army* seems an accurate representation! The Home Guard carried out numerous military training exercises, involving shooting contests and pub games with Leighton Buzzard Police. A bomb-throwing contest was organised in Pages Park on Boxing Day 1942. The men were trained in the use of the rifle, the sten gun and a whole series of other weapons that became available. These included the Ernest Youlle (EY) rifle grenade thrower, the Lewis gun, the Northover Projector, an anti-tank gun and the Spigot mortar, a mortar mounted to swivel round and fire in any direction. It was placed at strategic points, such as bridges, where the Home Guard stood in a circular trench for protection. The Home Guard was also trained in bayonet fighting, close armed combat, unarmed combat and camouflage.

A 29mm Spigot anti-tank mortar and base, which had a maximum range of 400yd. One of the stainless steel mounting pins used can still be seen outside Kimbolton Dental Surgery at the junction of Kimbolton Road (A428) and Goldington Road in Bedford. The rocket-propelled Spigot mortar was issued to the Home Guard in early 1942 but it was not popular and no more were issued after July 1942. Here we see the pit excavated round the concrete base to shelter the crew operating the mortar. This emplacement would have been built by the Works Service of the Royal Engineers.

One of the last combined military and civil exercises to take place at Leighton Buzzard, 'Vixen', on 10 January 1943, was curtailed due to bad weather: the icy conditions of the roads made it impossible for umpires and senior officials to arrive on time. The report in the *Leighton Buzzard Observer* of 12 January 1943 made no mention of the adverse weather conditions, but thought the combined exercise provided useful lessons for 'E' Company of the Home Guard, the ATC and the Boy Scout messengers, which seems to be a coded way of saying they were overwhelmed by the Regulars masquerading as enemy troops. The 500 Regulars, playing the role of airborne troops landing at Battlesden, dealt with three Home Guard 'heavily defended' outposts and filtered into Leighton Buzzard without suffering any 'casualties', possibly due to the shortage of umpires, and captured Cedars School after fighting in the grounds. 'The attackers generally gained their objectives by weight of numbers', said the report.

In 1943 a War Office Film Unit under Sergeant Parkinson produced a film showing battle exercise 'Spartan' in Leighton Buzzard and Oxford. The action was filmed from both the 'British' and 'German' sides. One of the scenes involved a

'German' sergeant pressing a plunger to ignite a charge and blow up a bridge near Leighton Buzzard before withdrawing. The troops for the film were provided by the 1st Canadian Division, the Guards Armoured Division, and the RAF Regiment, according to records in the Imperial War Museum. In July 1944 'K' Company of the 6th Battalion (Stanbridge and Eggington) took part in a training exercise finishing at Charity Farm in Eggington with a barrage from Smith guns and Spigot mortars.

The problem of deciding the most suitable size for a combined exercise involving both military and civilian services was never really solved. However, the personnel of the various services were generally agreed that an incident realistically staged with bomb damage and faked casualties was more helpful than imaginary incidents calling out a large number of services where the personnel had little practical work, although such an exercise gave valuable experience to control room staff. It is perhaps fortunate that the Bedfordshire organisation was never tested during the war.

The Women's Land Army

The Women's Land Army (WLA), created by Lady Gertrude Denman, who had chaired the National Federation of Women's Institutes since 1917, undertook a wide range of jobs in the war, from milking and general farm work to cutting down trees and working in sawmills as well as controlling pests. Volunteers had to be between 17 and 40 and of good physique; but in practice most were under 30. They lived in hostels or on the farms where they worked. Over the period 1939–50, over 200,000 Land Army girls, as they were generally known, served on the farms and market gardens of England and Wales. The work was hard and they worked long hours, especially during the summer, but without their efforts it is doubtful Britain would have had enough food to survive.

On 4 June 1942 the Leighton Buzzard hutment hostel became the second 'War Ag' WLA hostel in the county (after Milton Ernest, which was opened that February) in Hockliffe Road, where Appenine Way now branches off (Ordnance Survey SP93432543). It accommodated forty Land Army girls in two-tier bunks, thirty-nine from Bedfordshire and one from Flintshire.

The Linslade contingent of the WLA was accommodated at Harcourt House, a large private dwelling in Stoke Road, Linslade, not too far from 'Wilmead'. The house no longer exists, having been demolished after the war to make way for a new housing estate. *The London Gazette* of 16 February 1945 carries an entry concerning Jean Evelyn Tracey from Germany, who was living at Harcourt House. She is included in a 'List of Aliens to whom Certificates of Naturalization have been granted by the Secretary of State, and whose Oaths of Allegiance have been registered in the Home Office during the month of January, 1945',

WLA girls at the old LNER railway station in Bute Street, Luton, en route to Leighton Buzzard, 4 May 1942. Second from the left is Joyce How and third from the right is Ellen May Godfrey – these are the only women in the group that can now be identified by the authors.

The Leighton Buzzard Land Army girls in front of their hostel in Hockliffe Road, Leighton Buzzard. Since the recruits were supposed to be under 40 the older ladies are presumably support staff looking after the 'girls'.

in her case a readmission to British Nationality (she took the Oath of Allegiance on 25 December 1944). The major event recorded at the house was a four-day 'strike' beginning on 12 April 1945 in protest at the exclusion of WLA members from post-war benefits and privileges.

STELLA GOLDSMITH (NÉE LIMON)

On the 'People's Archive' website, Stella, a former clerk in an insurance company, indicates that she was one of the first Land Army girls to be accommodated in the Leighton Buzzard hostel:

When I applied to join the Land Army, Dad was not best pleased but being physically fit I was accepted and my uniform arrived. One felt hat and one overcoat, one green pullover, a pair of stiff corduroy breeches (which made a noise as they moved), two short sleeved shirts, two pairs of dungarees, two pairs of long khaki socks, a pair of brown leather shoes and a pair of black boots (which were agony until they were broken in), a red and green armband and a metal Land Army badge. So dressed in my newly acquired finery I set out for Leighton Buzzard. For some reason we were issued with a voucher for the railway – hardly the best way to get there from Luton.

We were stationed in a building on the outskirts of the town. A low brick building, T-shaped with the dormitory and ablution block running across the front of the building and the dining room, kitchen and warden's quarters leading back towards the 'garden' planted, all half acre of it, with Brussels sprouts. There were 40 of us new recruits. The dormitory was equipped with bunks made of plywood with a thin mattress on top, sheets, pillow and a couple of army type blankets. There was a single wardrobe between two occupants of the bunk. In the middle of the room were two iron stoves and a scuttle filled with coke. The ablution block had four toilets, four baths and six washbasins. You can imagine the rush after work to try and bag a bath before the water ran out. I shared a bunk with Rona, my colleague from the Prudential. I had the top bunk which to get into you had to climb on the chest of drawers.

It was all pretty basic … in the summer you would be kept awake with the sound of earwigs dropping on the floor and in winter the fire buckets kept by the stoves would be frozen solid every morning. The hostel had been built to house Italian prisoners of war but they had to be moved as the conditions didn't meet Geneva Convention requirements.

One issue that really upset the land girls was that they were taken to their farms in open lorries, even when it was cold and wet, whereas POWs were always transported in covered lorries. They put in a formal request for a canvas top to be put on their lorry and were promised action, but nothing happened. Finally, they decided that they had had enough and refused to go to work: instead they worked in the garden, weeding the Brussels sprouts. It wasn't long before a whole contingent of officials arrived from WLA HQ in Bedford – travelling by staff car – to threaten them with dire repercussions under the Regulations; but the land girls remained firm and cleverly continued working, so they could not be accused of striking. The next day a covered lorry arrived.

MARY 'MICKIE' HICKEY

Mary, a former window dresser in a boutique, was billeted at various hostels in Whipsnade, Kensworth and Leighton Buzzard. She stayed first at Whipsnade, which had been converted from a zoo to a farm. Only the parrots were left and they soon learned from the girls to swear: several of them would say 'bugger off' when someone came to the door. At Leighton Buzzard Mary picked potatoes, Brussels sprouts and did hay-making, amongst other things. She learned to smoke to reduce the smell of manure, which they had to roll up 'just like a carpet'. The hostel she liked most was Kensworth, a lovely old house altered to create more bedrooms and bathrooms. It had a library with huge windows and a legend of the headless horseman who used to ride by at night. One night, after coming back from a dance at Bedford Corn Exchange, she and a friend put white sheets over their heads and knocked on the door of a 'dear old lady' who had grassed them up to the warden about getting back late. Their 'ohhhhhhhhhhh' noises seem to have disturbed the 'poor old dear', as she left a week later.

MARY EMMA MARLEY (NÉE KENEFORD)

Mary recalled that on pay day, after working on Saturday morning, some of the WLA girls walked 3 miles (5km) into Hockliffe to a café where they had Welsh Rarebit, which was a real highlight, since food at the hostel was very basic – bread and jam or paste for 'bait' (packed lunch). They had a hot meal at night and porridge for breakfast, which she didn't like. But looking back, her Land Army days were the best of her life. She felt she had never lived till she joined the Land Army!

The hostel was closed in September 1950. No trace of it remains today. A full account of the WLA in Bedfordshire during the period 1939–1950 is given in *We Wouldn't Have Missed it for the World* by Stuart Antrobus.

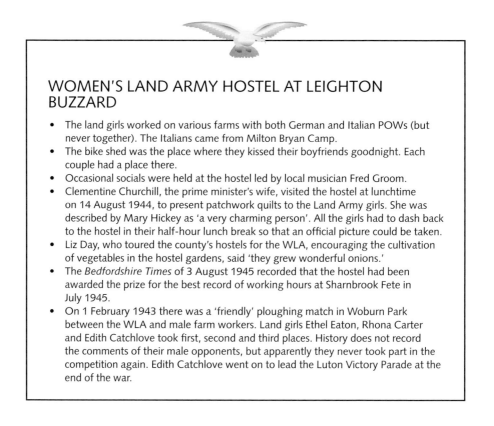

WOMEN'S LAND ARMY HOSTEL AT LEIGHTON BUZZARD

- The land girls worked on various farms with both German and Italian POWs (but never together). The Italians came from Milton Bryan Camp.
- The bike shed was the place where they kissed their boyfriends goodnight. Each couple had a place there.
- Occasional socials were held at the hostel led by local musician Fred Groom.
- Clementine Churchill, the prime minister's wife, visited the hostel at lunchtime on 14 August 1944, to present patchwork quilts to the Land Army girls. She was described by Mary Hickey as 'a very charming person'. All the girls had to dash back to the hostel in their half-hour lunch break so that an official picture could be taken.
- Liz Day, who toured the county's hostels for the WLA, encouraging the cultivation of vegetables in the hostel gardens, said 'they grew wonderful onions.'
- The *Bedfordshire Times* of 3 August 1945 recorded that the hostel had been awarded the prize for the best record of working hours at Sharnbrook Fete in July 1945.
- On 1 February 1943 there was a 'friendly' ploughing match in Woburn Park between the WLA and male farm workers. Land girls Ethel Eaton, Rhona Carter and Edith Catchlove took first, second and third places. History does not record the comments of their male opponents, but apparently they never took part in the competition again. Edith Catchlove went on to lead the Luton Victory Parade at the end of the war.

13

A TELEGRAM BOY: PERSONAL MEMOIRS OF JOHN VICKERS

These are the memories of another local man, John Vickers. He was a 14-year-old when he saw a post office messenger deliver a telegram to a coffee merchant opposite his home in 1942. He thought he would 'not mind doing that.' He spent the next three years delivering telegrams around the town, including to the secret establishments Q Central and No. 60 Group, without knowing how important they were. He still did not know seventy years later, when interviewed for this book:

I started on December 17, 1942 at Leighton Buzzard Post Office a week before my fourteenth birthday, where I joined two other boys on the staff (Horace Musty and Keith Audas) to be known by the Post Office title 'Boy Messenger' but more commonly known by the public as a 'Telegram Boy'.

When I started at the GPO I was asked if I would mind using my own bicycle as they hadn't a spare Post Office one at that time for me to use. I agreed to this, and was paid 4d a mile, but I was soon earning more than the more senior boys who were on a set weekly wage, so they quickly found me an official bike. Thereafter I earned the princely sum of fifteen shillings and eight pence a week. In due course I was supplied with a uniform – navy blue with red piping around the cuffs of the jacket and around the collar and edging, and down the seam of the trousers. This was similar to the postman's uniform, except that we wore a pillbox hat, also with red piping and a red button in the centre of the crown. We were also supplied with shoes for summer, and boots and a short cycling overcoat for winter together with a cycling cape and leggings for rainy days. I hated boots, but my mother said they gave my ankles the necessary support. When I was supplied with my uniform, I wore my own blue and white striped tie, as a tie was not supplied with the uniform, and was promptly told by the Head Postmaster that although it was rather a nice one, it was customary to wear a black one. I went out and bought one, and the next day my next-door neighbour asked me if I had lost someone in the family.

When we arrived on duty each day, we had to present ourselves to the Inspector for his daily inspection, and if our shoes or brass buckle and leather belt failed to attain the shine he desired, we were sent home to effect an improvement. I must have impressed him – I don't remember ever being sent home. Our bicycles had to be cleaned weekly, and these were also inspected to see that we had shone them well enough to be seen on the streets by the general public. I was not always so lucky at this – and was asked several times to try again. My request (with tongue in cheek) for a motorbike, brought a predictable response from the Inspector: 'The next thing we shall be getting you is a coffin.' Our appearance together with those of the postmen was of prime importance and we always had to appear to be seen to be correctly dressed. When I see the postmen of today and see how they dress whilst on delivery, I think of our old head postmaster who would 'turn in his grave' if he saw them. We would never be allowed to be out on the streets looking like that.

The messengers worked three different 'shifts' which started at 8.00 a.m., 8.30 a.m., and 10.00 a.m. and ending at varying times up to 8.00 p.m. We also worked a Sunday morning session, and also delivered to Wing on Thursday afternoons, and Heath and Reach on Saturday afternoons when their respective sub-offices closed for their half-day. We were at a slight disadvantage there as the locality and roads were completely foreign to us. When we were on the late shift, we used to go to the fish and chip shop to get supplies for all the staff back at the office who were also working late, and get various quantities of three 'pennorths' of chips and load them into our jacket in order to be able to ride our bicycles back. The leather belt around our waist prevented them from falling through. It was lovely and warm too. During the day, we also visited the local bakeries for iced buns and lardy cake etc.

Most of the postmen were elderly, as all able bodied men were in the armed forces, and we had quite a number of women both working in the sorting office and also as post women, to make up the required staffing numbers. We had our own detachment of the Home Guard and they used to parade outside the office in Church Square on Sunday mornings, and we used to hang out of our first floor window and 'take the mickey' out of them. They all took it in good part except one 'old timer' who used to get upset and say 'You can't do that to an old soldier'. I regret now that I never had a photograph taken of myself in my uniform, but films were rarely available at that time, as they were all required for the war effort, but I have a newspaper cutting of The Post Office Home Guard in Leighton Buzzard which brings back many happy memories.

The telegram service was eventually withdrawn as uneconomic, but was big business when I joined the GPO in 1942. Few people had telephones compared with the present day, and it was one of the quickest means of communication.

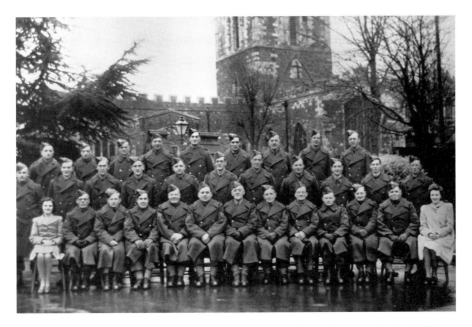

Many of the men who worked as postmen were too old to be called up so the Post Office formed its own Home Guard unit. Here the unit poses in Church Square outside the Post Office, with All Saints' church in the background.

The telegrams cost 9 words for 6d and a penny for each additional word. This included the address as well as the text, which resulted in very brief addresses and text to keep them within the minimum fee. Greetings telegrams were in a pale blue envelope with a specially designed form and were 6d extra. Priority telegrams, which were processed and delivered before all others, also attracted an additional fee of 6d. As our office was relatively small, and our weekly output of telegrams was about 500, we did not possess a printer for the telegrams but they were sent to us via a private line from Luton Post Office and the messages and envelopes were hand written by the staff on duty at our office. Once I delivered a telegram, and the recipient said on taking it from me, 'Oh it's a telegram from our Daisy.' 'How do you know that,' said her husband looking over her shoulder, 'You haven't opened it yet.' 'Oh, I recognised her hand writing on the envelope,' said the lady. I didn't think it would be proper for me to correct her.

The procedure on delivery was to hand them to the addressee, not merely push them through the letter box, and wait for the message to be read, and take any replies that they wished to send, and we carried spare forms for this purpose in our leather pouches. We counted up the words and asked for the appropriate payment. We also carried cards, which we left if there was no reply, indicating that we had attempted to deliver a telegram without success, and that we would try again later. Occasionally when I thought that the householder may soon return, I did put the telegram through the letter box and left a card at another entrance, usually the back door to show that I had left a telegram in this way. Once I left a telegram at the right house number, but I was in the wrong road (slapped wrist). On another occasion I had a lady call at my home for her telegram one evening as she recognised my signature on the card I had left. I suppose that she thought I took all the undelivered ones home with me when I finished duty.

On delivery I gave the doorknockers my usual rat-tat-tat. At one house I really went to town on the knocker, only to have it opened by the occupant who had been immediately behind it, with her hands over her ears. Door bells were another fascination and I used to occasionally ring them in the same manner as a telephone ringing, and was rewarded one day by hearing the lady of the house come into the hall of her home and picking up the telephone and saying 'Hallo'.

In our pouches we carried our rulebook, which stated that under penalty of disciplinary action we were not allowed to accept any gifts or gratuity. We were however, often given a tip of 6d or a shilling and sometimes if we were lucky, as much as two shillings. One day I was offered the princely sum of five shillings. This may not appear to be an unusually large tip, but in those days it was to me a third of my week's wages. The only trouble was that the generous benefactor was a postal official who would be bound to know the rules. It was the occasion of his daughter's wedding, and I was delivering some greetings telegrams – surely he wouldn't try to trap me on such a happy family occasion. I did not hesitate too long – he might change his mind. I left the reception a lot richer.

As well as the telegrams, we had to deliver Express letters and parcels, which attracted an extra fee of 6d in addition to the normal postage rate. These items were delivered as soon as they arrived in the sorting office, and the incoming mailbags carried a small extra label indicating that an express item was within the mailbag. This caused the bag to be opened and dealt with promptly, instead of waiting for the mail to be sorted and delivered in the usual manner. Our main customer for this express service locally was The United Molasses Company that had moved from their usual offices due to the war, although I can't remember where from, to occupy Sefton Lodge on the far side of Stockgrove Park. This firm had two Express items per day at about 10 o'clock and again at 3 o'clock. The normal way to get there was via Heath and Reach, and then via the Great Brickhill Road, but the easier and shorter way for us using a bicycle was to cycle through Stockgrove Park entering via Rushmere Lodge in Plantation Road. The park was private and we had to pass the home where the Kroyer-Kielbergs (a Danish family) who owned the park, lived. Because of the distance involved, we messengers took it in turns to do this longer trip. On my first visit, I was stopped by Mrs Kroyer-Kielberg, who told me that I was on private property. I explained who I was as I hadn't yet been supplied with my uniform, and she said 'oh, I often see you boys cycling through.' She allowed me to continue and I was never stopped again. It really was a lovely ride through the park with its rhododendrons and azaleas, and there used to be a lovely clock in a tower on an archway under which we cycled, which chimed the hours on some bells which hung on the outside of the tower. I understand that it no longer works, which is a shame – it was a lovely clock. The 51st Highland Division, which distinguished itself during the war, trained within the park. I subsequently cycled through there on numerous occasions in all weathers from lovely sunshine to snow blizzards. I remember going home to dinner one day from my ride through Stockgrove Park, and my mother brushed me over with a stiff broom to remove a thick layer of snow before I went indoors. I never pass Rushmere Lodge these days without remembering those rides through the park.

Delivering telegrams was not without its incidents – one local road, Broomhills Road, was notorious in my mind for its nosey women. They seemed to have in-built radar and as soon as the front wheel of my bicycle entered their road, heads would appear, just to find out who was receiving a telegram that day. Sometimes they would stop you and ask if you had one for them, knowing full well that I had not but they hoped that you would reveal which house you was calling at. I soon got wise to this ploy and determined to outwit them. The next time I went to their road I went via the passageway between the houses intending to call at the back door, so that the nosey ones looking out of their front doors would not know which house I was calling at. It was here I was promptly bitten by a dog for my troubles. A woman's head popped around the corner of the passageway to see what the hullabaloo was, and asked in a silly voice 'Oh dear, has he bitten you?' 'Yes, he flipping well has' was my reply. I duly reported it when I returned to the office, and my leg went on public display for all and sundry to observe the teeth marks and to comment. The inspector cheered me up no end when he said that if the dog's teeth had pierced skin he would have sent me to the doctor for a tetanus injection. The thought of that put me off from going down any other alleyways.

In 1943 a 500-lb high-explosive bomb fell in the courtyard of the old workhouse but the sandy soil prevented the bomb from doing much damage. It was my unhappy duty to deliver many sad messages. One I can still remember vividly was to a lady who on seeing me at the door when she opened it, took the telegram from me and danced a little jig in front of me, and called to someone at the rear of the house. 'Oh, it's a telegram from Bill – he is obviously coming home on leave and he has sent a telegram to say what train he will be on.' Alas it was not to be, it was a message from the War Office to tell her that he had been killed in action. On reading the message, the poor girl collapsed in front of me and sobbed her heart out. I was then only 14 or 15 years old and I hadn't witnessed an adult react quite like this before. I felt so helpless, and did not know what to do.

Fortunately war had its lighter moments. Leighton Buzzard had large contingents of the armed forces in town during the war, mostly RAF personnel. RAF Leighton Buzzard was completely under camouflage netting, and it was a strange experience to cycle through the camp, which we did several times a day. The road was completely blocked off to the general public, and when I went there on my first trip I was stopped at the gate by the sentry who said I couldn't pass. I was wearing a cape which we had for rainy days and also leggings due to the inclement weather, so I had to convince him that under this paraphernalia there existed a real telegram boy, even though I had not been supplied with my uniform, and only wore a temporary arm band to identify myself. I soon got recognised by them on my many subsequent visits, and I used to chat to the various guards on sentry duty. One, I remember used to tell me he worked in a film studio before being called up, and he used to thrill me with his tales of life with the stars. Whether it was all true or not, I don't know, but it remains a happy memory of those days.

There were many smaller contingents in town in all sorts of commandeered buildings including a section of the WAAFs in a house opposite the Post Office in Church Square. This house had been requisitioned for their use and we had a daily view from our messenger's room of the girls either arriving off duty and getting undressed and going to bed, or getting up and getting dressed to go on duty, due to the lack of curtains on their windows. This lasted for some time until the girls wrote home and asked mum to send them something to cover the windows, but there were some who never bothered. One day I had a telegram to deliver to this house. The front door always stood open and as I couldn't find anyone about, I ventured in and studied the names on the doors. I found the name of the girl I had a telegram for on the first floor, so I knocked the door. 'Come in' said a sweet voice. I ventured in and found the girl in bed. I don't know who was the most surprised, – as I am sure she wasn't expecting a male visitor, neither did I expect to find her as the sole occupant of the room in bed. When she read her telegram she said: 'you are one of the boys who watch us getting undressed each day aren't you?' I then began to wish that I was watching from the safety of the room across the road.

One of the chores we had to do was to sort out mailbag labels, which in wartime we had to keep reusing until they were no longer serviceable. We tied them up into bundles and returned them to the originating office for reuse. With these, we had to sort out the lead seals which were used to seal the mailbags and on removing all trace of string, bag them up to be melted down to make new seals. The worst chore was the string – this we had to untie and make bundles of it in suitable lengths for reuse. Those knots. It took a great deal of patience to untie the string and the piles of it got ever higher. One day I found a loose floorboard in our room and promptly secreted it all away out of sight. If anyone should lift those floor boards on the first floor in Leighton Buzzard Post Office now, they might be led to believe that they had found some wartime expedient for under floor insulation – if so, I have news for them.

14

BARRAGE BALLOONS, BRAS AND SECOND-HAND CARS

Companies participating in the war effort included Coty, who produced flares, R.F. Hunter in Vimy Road (originally of Gower Street in London WC1), who made cine screens for the military, and Cotsmoor, who made parachutes. One of the country's most famous makers of brassieres and corsets both before and after the war, Gossard turned their workers' skills from 1939 to 1945 to saving the lives of Allied servicemen and women. They made dinghies for fighter aircraft so pilots who ditched in the sea could escape, thousands of barrage balloons for convoy ships so attacking dive-bombers would be tangled in the wires, and vast numbers of parachutes.

Ironically, Gossard's factory in Grovebury Road had been built in the First World War by the government for the manufacture of torpedo netting to protect ships from underwater attack by U-boats. It was known as the wire works but in 1926 was converted for making 'foundation garments'. When the war broke out the payroll was 180, but this was doubled when the Ministry of Aircraft Production asked the factory to make war equipment. The existing machinery was adapted for parachute manufacture and additions were made for other needs.

During the war the Gossard works in Leighton Buzzard employed 400 and another 30 worked in Linslade. The company made 5,000 convoy balloons, 100,000 single-seater fighter dinghies, 639,309 parachutes of all types, 19,000 lifebelts, 35,000 distress flares, 73,500 sails, 348 experimental balloons, 30,000 man dropper parachute repairs – and, keeping old skills going, 120,000 Wrens' brassieres. Vivid pictures of their equipment in action were placed around the walls of the factory to provide 'stimulus to the worker'.

At the end of the war, when a new social club was opened for the workers, there was an overwhelming demand for new brassieres. The staff had shrunk to 170 because many of the workers had been recruited from other towns and returned home, so 140 new workers were needed to meet demand.

Barrage balloon production at one of Gossard's premises in Leighton Buzzard in 1942. This would have been unpleasant work using noxious solvents without the protective measures that would nowadays be in place. The kneeling figure in the white blouse is Irene Ruby Hart of Old Road, Linslade. Her married name was Orchard; she died in April 2013 at the age of 93. (Collection of Harry Maughan. Mrs Orchard was the aunt of his late wife, Frances.)

Part of the war effort: the bronze disc on the 'Fly Past' monument at the junction of West Street and Bridge Street in Leighton Buzzard, unveiled in September 2005, shows 'bras and parachutes', two of Gossard's products.

Coty's workforce, who made pathfinder and reconnaissance flares based in prefabs in the fields of the Rothschilds' stud at Southcott. Another field was used to test the gunpowder. Among the workers was Yvonne de Rothschild, wife of Anthony, and mother of Sir Evelyn de Rothschild, who still lives nearby at Ascott House, Wing. The picture, taken in June 1945, shows, among others, Thelma Seed, Miss Bates, Margaret Rickard, Mrs Andrews, Iris Sage, Mrs Palmer, Joan Hall, Edith Thorpe, Margaret Randall, Harry Corkett, Georgina Dimmock, Mr and Mrs Albert Dimmock, Olive Farmborough and Ray Probert.

Part of the Coty's workforce who worked in the High Street, Leighton Buzzard, checking and folding the parachutes for the larger pathfinder flares. These were fired off once a week from Southcott to check they worked properly.

The Cotsmoor clothing factory in Leighton Buzzard was easily adapted to sewing parachutes. Its staff had all the necessary skills and equipment and turned out thousands during the war.

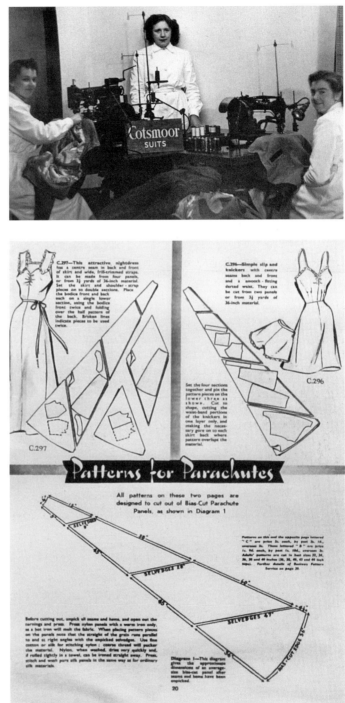

With the rationing of clothing and dressmaking material, any 'points-free' fabric was treasure indeed. Since parachute fabric had been cut on the bias, some ingenuity was called for. The instructions for making clothes out of parachutes included a warning not to use a hot iron on the nylon. This being a newly invented material, most people would not realise a hot iron would melt it.

One of the tributes paid to the 'girls', still on the wall at the opening of the club, was from the crew of the Dutch vessel HNLMS *Juliana*:

> The Germans hate and fear your work but we appreciate it. We look on your balloons as a definite and very effective part of our ship. You should all be cheered by the number of Jerries who had got mixed up with the balloons and been sent crashing into the seas. The number of ships which have been definitely saved is really encouraging.

The Michaels Brothers and Camden Motors

To avoid the expected bombing of London two brothers – Henry and John Michaels – came to Leighton Buzzard at the start of the war and set up a second-hand car business in Lake Street. They called it Camden Motors after the home they had left. They were Jewish, possibly, about 40 – too old for the draft – and they came with their families. One of them built a house opposite the car showroom next to Perth House; the other lived further down Billington Road. The car showroom became the biggest second-hand car dealers in the south of England and attracted people from many miles away.

Bill Marshall, a Leighton Buzzard teenager during the war, worked for the Michaels as a farm labourer. Under wartime regulations all land had to be put to the plough so the meadow behind the showroom, later used to store cars, was used to grow wheat. Bill helped with the harvest, gathering stooks (of cut wheat) together. He was happy to work for them because they paid more than most, 1*s* an hour, good wages for the time.

The second-hand car business was originally founded in 1932 by Henry Michaels but came to Leighton in 1939 with a dozen old cars. It expanded rapidly after the war, selling new as well as used cars. By 2012 the original site in Lake Street had been closed down and sold off for housing, but the company still retained its head office in Ridgeway Court, Leighton Buzzard, and had thirty franchises and 1,600 employees.

15

HELPING WITH THE WAR EFFORT

Life in the war years was very different from anything we are familiar with today. It was total war, involving the general public in immense efforts as well as the military. There were no supermarkets or public television. Everyone was required to carry a registration card, which was linked to the issue of a ration card. Children wore an identity bracelet with their name and number on it. The recommended level of water for a bath was 5in. You could be fined for leaving lights on in an empty house, not observing the blackout, or for putting bread on a bird table. Wardens hammered on doors to enforce the blackout. You were encouraged to grow tomatoes in window boxes, raid the hedgerows for blackberries, plant vegetables in your garden and eat everything on your plate, and you were told you were lucky to get it. The black humour of 'Hello meat' pies and dumplings meant they consisted mostly of vegetables. Worst of all, public houses sometimes had to put up the sign 'No Beer'. Signposts were taken down and traffic islands and walls on bends painted white to prevent accidents. Gas masks of all sizes were issued and also small pieces of solid rubber to chew on in case of bomb-blast. Some dogs were provided with gas-proof kennels. You learnt not to throw anything away. The national watchword was 'use it up, wear it out, make it do, or do without.'

In December 1941 the Ministry of Food issued guidance on children's party food. In its advertisement 'Let's talk about Xmas Food', it suggested simple treats such as gingerbread men cut from pastry or biscuit dough, cakes iced with a mixture of cocoa, sugar and milk, and open sandwiches of raw vegetables with perhaps a sardine and a little cheese. For Christmas decorations, it recommended holly dipped in a strong solution of Epsom salts, which dried to give a frosty sparkle. And for a blaze of colour it came up with the idea of a jolly bowl of vegetables: 'the cheerful glow of carrot, the rich crimson of beetroot, and the emerald of parsley'. Clearly the Ministry's officials were doing their best to put on a brave show, but it's as well that children had less sophisticated tastes in 1941 than they do today.

The extra psychological stress of wartime was countered by *Woman's Own* magazine in 1940 with suggestions for treating headaches, constipation and a feeling of being generally run down. Ovaltine and Horlicks were regarded as particularly good for sound sleep. But on the whole the British people were cheerful, warmly clad and far from starving. You actually talked to your neighbours and perhaps did a bit of bartering with them. You got healthy exercise by walking a lot and driving very little.

The stories below indicate something of the effort that people in and around Leighton Buzzard put into their many and various contributions to winning the war.

Fundraising, Salvage, a Spitfire, a Warship and a Tank

One of many salvage drives was held in Bedfordshire and Buckinghamshire in June 1942, but in Leighton Buzzard an intensified drive got underway and the streets were systematically canvassed. Linslade too made a special effort. Leighton Buzzard's opening ceremony commenced in an unusual way – W.D. Cook, Chairman of the Urban District Council, drove a steam roller up the High Street crushing piles of tins because this made them easier to transport to the foundries. In the Market Square was a display of various weapons and articles of war made from salvaged metal. The shop windows had specimens of the salvage required, which included bones, scrap iron and metals, rubber, bottles, tins, rags and textiles. The importance of bones was particularly emphasised, as too many were wasted in Leighton Buzzard. It was calculated that a hundredweight (cwt) per 1,000 population was not accounted for every month, even after allowances were made for bones given to dogs and the amount of boneless meat coming into the town. Bones were used to make glue for aircraft production.

Without salvage there could be no munitions: it was the one good source of raw materials. Each person was urged to do his or her bit to save waste materials and not leave it to their neighbours. People were told by Mr Cook that the shipping situation was serious and vessels could not be spared to bring raw materials to Britain.

The following week a mile of books was requested in Leighton Buzzard's High Street, which certainly caught the public's attention. Starting outside Messrs W.D. Cook & Sons' shop, books were laid along the pavement on the roadside to the council offices in North Street; thence back to the Roebuck corner on the other side; and from there to Dudley Street via Lake Street and then back to the Post Office in Church Square. WVS and Girl Guides acted as stewards. The 'mile' was started with Dickens' *Bleak House* and the books discarded ranged from old-fashioned volumes to recent popular fiction, geographical magazines, old lesson books and ledgers. A volume of *Great Victorians* lay next to four well-bound volumes of *The Connoisseur*, Bunyan's *Pilgrim's Progress*, old medical and legal tomes, and music, poems, etc. Occasionally there were books whose value was not realised and it was pointed out that two volumes were worth £20. Dr William Bradbrooke's *History of Fenny Stratford*, published in 1911, was one of only thirty-seven copies ever printed. Exactly three-quarters of a mile of pavement were covered, but if the books had been laid end-to-end more than a mile would have been covered. A baling machine was set up in the Market Square and was exceptionally busy during the morning. Children and Boy Scouts proudly brought trucks filled with old books and waste paper and thoroughly enjoyed this part of the war effort.

In January 1942 the biggest waste paper campaign ever launched was held to provide war factories with essential minerals. The Waste Paper Recovery Association provided £20,000 in prize money, of which £1,000 was given to each of twenty defined areas. These prizes were divided among the local authorities in each area (comprising city, urban or rural districts) that had collected the greatest weight of waste paper and cardboard per head of population by 31 January, based on the returns of the local Food Executive Officer.

The £1,000 was divided into awards of £500, £250 and £100, and three consolation prizes of £50 each. One half of any prize money went to either the Red Cross, Mrs Churchill's Aid to Russia Fund, the RAF Benevolent Fund or the Soldiers', Sailors' and Airmen's Families Association and the other half to local charities.

Many people felt at first that they had little paper to offer, as the daily and weekly newspapers were needed for fire lighting and shopping purposes. But they soon realised that they could contribute by collecting tea wrappers, biscuit containers, chocolate papers, empty cardboard boxes, etc. Leighton Buzzard collected 39 tons and 18cwt (915kg) of paper. Of this total, 28 tons 15cwt was collected from houses and shops. The remainder was credited to the town by firms who sent their waste direct to mills or merchants. The rationed population was 8,166 people and the town saved nearly 11lb (5kg) per head. The normal monthly collection in Leighton Buzzard was 1 ton per 1,000 population.

Another book drive in July 1943 in Leighton Buzzard collected over 15,000 books. One older gentleman brought along a huge family album of photographs. 'What can you do with these? he demanded. When it was explained to him that they could only be used for pulping, he visibly brightened up. 'Good', he exclaimed, 'The photos are of my wife's relatives.'

Among other activities, individuals, and sometimes the whole of the town and surrounding villages, embarked on fundraising for particular armaments. In 1942 Gossard's employees set out to raise the cost of a Lewis gun (£75) and actually raised £88 5s. The first major project was for a town Spitfire, the second a warship and the third a tank. Below are the stories of the Spitfire and the warship,

HMS *Oakley*. All we know of the tank is what the *Leighton Buzzard Observer* of 28 September 1943 recorded:

> A lieutenant and four members of the King's Royal Hussars have sent a letter to Mr W S Higgs, Chairman of Leighton Buzzard Urban District Council, saying that they have received a new tank provided by the people of Leighton Buzzard and district, and thanking the district on behalf of the crew for this magnificent contribution to the war effort. They feel sure that the plate which is fixed to the front of the tank will bring them all good luck. The letter is signed by Lieutenant M Mather, Driver C Stephenson, Wireless Operator, G Miles, Gunner H Clegg and Forward Gunner J Nicholson.

The Spitfire Fund

In late July 1940 an appeal to the people of Leighton Buzzard, Linslade and Wing was mounted to raise enough money to present to the Ministry of Aircraft Production to buy a Spitfire fighter plane. A novel mock-auction sale 'of one whole Spitfire in separate parts' was held in Leighton Buzzard market on 2 September 1940. An RAF squadron leader listed all the parts of a Spitfire on cards, together with an estimated cost, and the cards were handed to purchasers as mementoes. Additional contributions came from the sale of Spitfire badges, a popularity contest and the sale of Mr Ivester Lloyd's hunting picture (£239). The money raised during the Spitfire Fund campaign was shown on the giant barometer in Church Square, as was later the case with HMS *Oakley*. Some of the money was raised by charging people to look at the wreckage of a German Messerschmitt on show in the grounds of Cedars School. Over £1,500 was collected in the first five days and by 18 February 1941 the appeal was just £500 short of the required total of £5,000, exclusive of expenses. To make up the remainder, the Fund Committee decided to invite 800 people to subscribe 10s (50p) each.

The Fund was closed in June 1941 when the total of £5,000 was achieved. A cheque for £2,500 was sent to the Ministry of Aircraft Production (a similar amount having been sent previously), with a request that the Spitfire be named 'The Buzzard'. And so it was. A response from John Moore-Brabazon of the Ministry to Mr W.S. Higgs, the Chairman of the Fund, expressed deep appreciation and said he would certainly arrange for the aircraft to be named as required.

The Buzzard fought with RAF 19 Fighter Squadron based in East Anglia and was used on convoy protection duties along the East Coast for most of 1942. It was damaged during an unrecorded air battle, but the pilot managed to

The Buzzard, a Spitfire VB with Merlin 32 engine, RAD No. AD556, presented to the Ministry of Aircraft Production by Leighton Buzzard, Linslade, and the Rural District of Wing during War Weapons Week, 1941.

land the plane and it was repaired. Later in the war the Buzzard was converted to a photo-reconnaissance plane. Extra fuel tanks and camera equipment were added and the three-blade propeller was replaced by a five-bladed version more suitable for high-altitude work. The Buzzard served out the rest of the war in this capacity, providing vital information for D-Day and the drive on Berlin. It lasted the war out, being scrapped in 1945. It had been an unusually long career for a combat aircraft.

HMS *Oakley*

In March 1942 a 'Warship Week' was mounted by Leighton Buzzard and Linslade Urban District Councils and Wing Rural District Council. The target was to raise £250,000 to purchase the destroyer HMS *Oakley* outright but, more realistically, if £150,000 (the cost of the hull) could be raised the ship would be 'adopted' and plaques exchanged. The ship was named after the village and civil parish of Oakley in northern Bedfordshire.

The inaugural parade for the week, described in the *Leighton Buzzard Observer* for 17 March 1942, was impressive. The Bedfordshire and Hertfordshire

Regimental Band headed the procession and a Czechoslovak army band and Leighton Buzzard Army Band also gave their services. Assembling in New Road, Linslade, the units were marshalled by Warrant Officer Norman RAF and the salute was given at both Linslade and Leighton Buzzard War Memorials during the march past. Lieutenant-Colonel C.M. Parkin, RA, mounted on horseback, was officer commanding the parade, and he was followed by RA units, the King's Royal Rifles, B Company of the 3rd Bedfordshire Home Guard (under Major T. Crockett and Capt. L.F. Vick), C Company 1st Buckinghamshire Home Guard, a Czechoslovak unit, and contingents from the RAF, WAAF, ATC, Bedfordshire and Buckinghamshire Police, the war reserve and special constables, Leighton Buzzard and Linslade ARP, the St John's Ambulance Brigade, the NFS and the Girl Guides. Two armoured cars also joined the parade.

In Leighton Buzzard recreation ground Sergeant Kenna acted as marshal and Lieutenant-Colonel Parkin gave the orders. The Union Jack was carried by a Chelsea pensioner and the National Anthem was sung. Admiral Sir Lionel Halsey gave an address, reported by the *Leighton Buzzard Observer*, in which he said that for two-and-a-half years the Royal Navy, Merchant Navy and British fishing boats had been fighting, and now that the war had extended to the Far East their job had never before been equalled in British history. Ships of all sorts and sizes were needed because the Japanese had temporarily wrested command of the seas in the Far East.

As with all wars there was belittling of the enemy, as the quoted remarks from W.D. Cook, Chairman of Leighton Buzzard Urban District Council, show. He was presiding over the whole event and reminded the large company of the target figure and the hope of raising £400,000: 'The district had done well in a previous effort by raising £5,000 for a Spitfire', he said. 'It was now the turn of the Navy. The personnel were there and if given the ships they would give those ugly little devils, the Japs, just what they had been asking for.'

Supporting Cook and Sir Lionel were officials of the Czechoslovak Army and Mrs Beneš (wife of the president), Colonel Charles Calveley Foss, VC, Rev. C.W. Morris (Chairman of Wing Council), Mr J. Page (Chairman of Linslade Urban District Council), Messrs F.S. Parrot, Valentine Charles Pool, A. Vickers, Wilfred Ernest Durrell, L.B. Faulkner, William Allder, S.R. Dillamore, Rev. A.T. Stephens, L.E. Lydekker and Oscar Watson (Hon. Secretary).

The parade afterwards marched along Grove Road, Lake Street and into the High Street, where Major-General Lionel Hugh K. Finch, CB, DSO, took the salute. After the parade the principal people in the procession accepted the invitation of Mr and Mrs W.D. Cook of Shenley House to toast HMS *Oakley*. A telegram expressing good wishes and stressing the need of support was received from Mr Albert Alexander (First Lord of the Admiralty):

I send you my best wishes for the success of your Warship Week. The money you lend means to the Admiralty ships, aircraft, guns, torpedoes and mines, and we need all we can get. The extension of the war has put a tremendous additional strain on the Royal Navy. Our sailors are standing up to it magnificently. I am sure you will see to it that the funds raised are not lacking to supply their needs. Whatever amount you first thought of please double or treble it. That will be the surest way to victory.

Sir Howard Kingsley Wood (Chancellor of the Exchequer) also telegraphed: 'I send all good wishes for Leighton Buzzard Warship Week. At no time has it been more vital that we should continue to stand four square behind our fighting men. Let this week's results reflect your determination to play your part.'

One of the first investors was young Anthony Gotzheim of Queen Street, who stood outside the Gas Company centre from 8 to 10 a.m. until they opened

The start of war savings week in August 1942. The barometer, fixed to the railings of Cedars School in Church Square, shows a target of £250,000 to help buy HMS *Oakley*. It was launched by local MP Alan Lennox Boyd.

to collect donations. Tradespeople decorated their premises and many windows contained models of ships. The appeal got off to a very good start, the first day's investments alone totalling £50,735.

A giant barometer was erected in Church Square outside Cedars School (as it then was), and the money raised was marked by a red line which moved up steadily, thanks to the generosity of the local people. A sum of £161,000 in loans and gifts, equal to £7 8s 4d per head of the local population, was raised towards the construction of the £400,000 ship. The gifts included £336 from the staff and pupils at Cedars School towards the cost of the ship's anchor, the proceeds of a dance at the Corn Exchange, at which Lawrence Inns' band played, and a contribution from Pitstone's Entertainment Fund for a machine gun. Sir Howard Kingsley Wood sent a second telegram: 'Heartiest congratulations on success of Leighton Buzzard Warship Week.'

In December 1942 a Comforts Fund Committee was set up by the three councils (Leighton Buzzard Urban District Council, Linslade Urban District Council and Wing Rural District Council) to assume responsibility for the well-being of the crew of HMS *Oakley* – at least to the extent of providing cigarettes, tobacco, games and books. Their first action was to send a cheque for £25 to the captain to spend on his crew at Christmas.

Local notables, including the town GP, Dr Square, from Bridge Street; officers from the services; and a group of Chelsea pensioners who had been evacuated to Ascott House, Wing, are addressed by Alan Lennox Boyd MP at the Leighton Buzzard Recreation Ground at the start of fundraising for HMS *Oakley*.

HMS *Oakley* Goes to Sea

In the 'Chronological Record of War Service of HMS *Oakley*', held in the Leighton Buzzard Library, Lieutenant Commander Geoffrey B. Mason describes the war record of the ship.

HMS *Oakley* was a Type II Hunt-Class Escort Destroyer ordered from Yarrow, Scotstoun, Glasgow under the 1939 War Emergency Programme on 20 December 1939. The ship was laid down on 19 August 1940 as Job No. 1853 and was to have been named HMS *Tickham*. However in June 1941, after a sister ship launched as HMS *Oakley* had been transferred to Poland, she was launched as the third HMS *Oakley* on 15 January 1942. Completion was delayed until 7 May 1942 because the shipyard was bombed before the hull was launched. Further misfortune occurred later in her career when the ship was attacked twice by British Spitfires that mistook her for an Italian torpedo boat.

HMS Oakley (L98) served in the Arctic in 1942 and off Sicily in 1943, supported the landings in southern France in 1944, and escorted convoys in the North Sea in 1945, gaining battle honours for all these theatres of operation. In total, she escorted twenty-two convoys during the war.

After the war, HMS *Oakley* was given a complete refit at Taranto in southern Italy and in October 1945 she returned to the UK, where she was placed in Reserve 'Category A' Service at Portsmouth. Ironically, she was sold to the Federal German Republic in November 1957 as a gunnery training ship and

HMS *Oakley*, a Type II Hunt-Class Escort Destroyer. (© IWM A25403)

The badge of HMS *Oakley*.

The commemorative plaque on the badge of HMS *Oakley*.

renamed the *Gneisenau* after the German battlecruiser that had been such a threat in the war. She served in this capacity until January 1977, when she was sold for breaking up.

All that remained in the UK was the ship's badge, a heavy cast-iron shield bearing a bugle horn erect and an annulet interlaced gold on a red field. This was discovered in 1975 by the town clerk, Peter Quick, during a clear-out of cupboards at The White House, the Leighton-Linslade Council Offices in Hockliffe Street, Leighton Buzzard. After a search in 2013, it was located again and is now held in their basement, a suitable item for a Leighton Buzzard Museum, should one ever come into being.

Recreation

With so many troops stationed in the Leighton Buzzard area it was important to provide entertainment for both them and the local population. Unlike many military towns there were more young service women than men. There were half a dozen dance halls providing live bands as well as three cinemas. Films at the Oriel cinema featured well-known stars including Betty Grable, Mickey Rooney, James Cagney, Arthur Askey, Bob Hope and Bette Davis. The programme list for April 1944, price 1*d*, included *Lost in a Harem* with Abbott and Costello, *Shine on Harvest Moon* with Ann Sheridan and Dennis Morgan, *Madonna of the Seven Moons* with Stewart Granger, Phyllis Calvert and Patricia Roc, *Practically Yours* with Claudette Colbert and Fred MacMurray, *He Stoops to Conquer* with George Formby, *The Big Noise* with Laurel and Hardy and *Once upon a Honeymoon* with Ginger Rogers and Cary Grant. Prices ranged from 10*d*, 1*s* 6*d*, 1*s* 9*d* and 2*s* 3*d* to 3*s* 6*d* for the really well off.

The cinema also showed propaganda and Ministry of Information films. One of the latter, shown at the Oriel on 8 June 1941, was entitled *Britain at Arms*, believed to have featured RAF officer Kenneth Douglas Faulkner, whose parents lived in Stoke Road in Linslade. He won the DFC and was killed in a flying accident in 1945. One very successful propaganda film was *The Lion has Wings* about the RAF, produced by Alexander Korda on the outbreak of war and starring Ralph Richardson. Another, *Target for Tonight*, was a 1941 British documentary filmed and acted by RAF personnel, all while under fire. Directed by Harry Watt, the film follows the crew of a Wellington aircraft on a mission over Germany. It went on to win an honorary Academy Award in 1942 and 'Best Documentary' from the National Board of Review in 1941. Talks about the war were also held at cinemas and other locations. One such was 'Daily Life under the Nazis' by Frau Litten, given on 22 November 1943, at the Unionist Club in the High Street, Leighton Buzzard. Admission was free.

Dances were held most evenings, Forces personnel in uniform paying about sixpence less for entry (normally 2s 6d). The regular dance bandleaders, including Lawrence Inns, Syd Fletcher, Fred Groom and Austin Farrell, played at the Co-op Hall in the High Street, the 'Old Vic' in Hartwell Grove, the Corn Exchange in Lake Street and the Village Hall, Wing. Eileen Younghusband records that, despite being very tired during her intensive training course at Oxendon House as a WAAF Clerk Special Duties for RAF 60 Group, she found the energy to visit the NAAFI for a bit of relaxation and go to dances on a Saturday evening with an RAF band, the Oxendonians. A crowd of airmen from a nearby operational fighter station were invited to the dances. The WAAFs were warned not to discuss their training with anyone so they forgot about plotting (at least as far as aircraft were concerned) and danced the night away.

Stella Goldsmith (née Limon) of the WLA says:

> Every so often there were dances in the WLA hostel (on Hockliffe Road) when members of the RAF stationed near Leighton Buzzard would be invited. They used to be great fun and a welcome break from routine. One drawback would be food. In order to provide refreshments at the dance, the ingredients would be taken from our ordinary rations for about a month beforehand. This, of course, meant even plainer food and no treats. Still, come the night of the dance and the fight to have a bath and a place by the wash basin, plus the chance to wear 'civvies', we all agreed it was all well worth it. It was quite a novelty to dance to gramophone music as other dance venues always had a live band.

Other entertainment included whist drives, farming demonstrations, gymkhanas and the occasional fair at Bridge Meadow, plus recitals by the four Croxford sisters whose father owned the menswear shop in town. In February 1940 a committee under the chairmanship of Rev. A. T. Stephens recommended the establishment of a social club to cater for members of the armed forces. With the support of the County Welfare Office, the centre was established in the Temperance Hall in Lake Street and equipped with a radiogram, dartboards, card tables, board games, crockery, magazines and writing materials. A canteen was run by members of the local WVS branch, who also offered to do mending for male members of the forces. Within a month the centre was on a sound financial basis and being increasingly used, although the demand to take up the mending service proved very small.

From 1942 the public was encouraged to spend their 'Holidays at Home'. Leighton/Linslade Chamber of Trade suggested all retail traders except those providing food and medicine should close to give their staff a much-needed week's rest in August. Public activities during the week were to include American tennis, a baby show, cricket match, angling contest, darts and boxing tournaments,

The delights of 'Holidays
at Home'.

LEIGHTON BUZZARD AND LINSLADE

HOLIDAY AT HOME WEEK.

AUGUST 1 to AUGUST 7, on BELL CLOSE.

SUNDAY:	OPEN-AIR SERVICE.
MONDAY:	ADULT SPORTS.
TUESDAY:	CHILDREN'S SPORTS.
WEDNESDAY:	N.F.S. DEMONSTRATION and Children's DANCING DISPLAYS.
THURSDAY:	COMIC CRICKET MATCH, DARTS TOURNAMENT. CONCERT at Corn Exchange
FRIDAY:	SHEEP DOG TRIALS (Admission 1/-). BOXING TOURNAMENT.
SATURDAY:	CRICKET MATCH. BABY SHOW. GRAND CARNIVAL.

OPEN-AIR DANCES each evening except Thursday, 6d. extra.

WHIST DRIVES and BOWLS TOURNAMENTS.

FUN FAIR DAILY.

Admission: Adults 6d., Children 3d.

Programmes on sale shortly.

children's sports, sheep dog trials, dances, concerts and a road cycle race to Aylesbury and back.

In 1943 the County War Agricultural Committee arranged a Utility Holiday, which provided additional labour for farmers and a healthy holiday for campers. The working day was 8 a.m. to 5 p.m. with optional overtime and a lunch break in the field. Each camper had to bring ration card, plus towel and soap; and a wide variety of meals would be provided. The pay was 1s for men and tenpence for women, teenagers receiving a suitable remuneration.

The local Grand Union Canal was used for outings and trips. On one occasion over 200 people from the Baptist chapel in Hockliffe Street took a canal trip on Faulkner's local sand barges to Slapton where races and games were arranged.

Whipsnade Zoo, 7 miles away, was a popular attraction, with over 6,000 visitors over Easter 1944 and many hundreds used cycles, tandems and tricycles to make the journey.

Salvation Army: Miss Ethel Quick

Miss Ethel Quick lived at 98 Hockliffe Road and suffered rheumatoid arthritis, which confined her to a wheelchair. She could not move her fingers, but she could wield a pen between thumb and forefinger and produced beautiful calligraphy. She was no slow motion writer either, in spite of her tremendous handicap. She, along with the rest of her family, was a staunch member of the Salvation Army and a Sunday School teacher for twenty years.

In 1940, when Ethel was 32, Major E. Wood (Salvation Army) suggested that she might like to write to the local Salvationists who were being called up to give them all the Salvation Army news. Ethel took up the offer, saying 'I decided to take on the job of Salvation Army war correspondent to the forces.' For a long time she kept up communications with fourteen members – twelve men and two women – serving in various theatres of war and altogether wrote them 1,056 letters and received over 500 replies. Even with her health problems, Ethel sometimes wrote up to twelve letters a day telling the men and women all the news concerning the local Salvation Army Corps activities, about the band, songsters and the different items at the meetings.

This lady's amazing effort did not go unnoticed. Ethel received a letter of congratulations from the queen, which was presented to her with other congratulatory messages at a Salvation Army victory tea and concert. Signed by Her Majesty's lady-in-waiting, Kathleen Seymour, the letter read:

> I am commanded by the Queen to write and say that her Majesty has heard of all that you have done to help the cause during the past sad years of war, and her Majesty was, indeed, interested to know that you have been able to write over 1,000 letters to local servicemen in spite of the fact that you are, alas, suffering from rheumatoid arthritis. The Queen does, indeed, admire the spirit which has enabled you to overcome such very great difficulties, and Her Majesty congratulates you on all that you have achieved in this time of trial.

Another touching moment during the proceedings was the presentation of a bouquet of chrysanthemums from five children whose fathers had received many of her letters. Attached to the flowers (grown by Mr T. Janes of Dudley Street) was a card inscribed: 'To Ethel, with love for remembering our Daddies – Brian Janes, Brian Garner, Colin and Barry Bray and Jennifer Aris.'

This all came as a complete surprise as the ceremony was a well-kept secret organised by Corps Sergeant-Major J. Tyler. So when Ethel was wheeled in her chair to the function she had no notion of what was going to happen. The presentation of the queen's letter, in a sealed envelope, along with a telegram, was made by Mr Tyler at the ceremony immediately preceding the concert, presided over by Miss Amy North, WAAF, who later read out the royal message. Also present were Major and Mrs Portus, officers in charge of the local corps of the Salvation Army, Major E. Wood and Adjutant White (corps officers here at the beginning of the war and later at Teddington) and Majors McBride and Nicholson of London, who were visiting the local corps.

Ethel admitted that she was thrilled and excited and the news was met with thunderous applause from the 100 or more who were present. In a very brief speech of acknowledgement she thanked especially her father (Mr A.J. Quick) and her brothers Horace and Albert and all those who had helped her to go to meetings, making it possible to send letters (from the *Leighton Buzzard Observer*, 13 November 1945).

16

FOOD RATIONING AND THE KITCHEN FRONT

In 1940 wasting food was a criminal offence. Everyone received their ration book on 1 January 1940 and food rationing came into effect a week later. The food ration varied throughout the war according to the availability of each commodity. Initially only bacon, butter and sugar were rationed, but as the war progressed more foods were added.

This was the standard ration for an adult per week:

Bacon/Ham	4oz (100g)
Meat	to the value of 1s 2d (6p in today's money).
Butter	2oz (50g) same coupons as margarine
Cheese	2oz (50g): sometimes it rose to 4oz (100g) and even up to 8oz (225g)
Margarine	4oz (100g) same coupons as for butter
Cooking fat	4oz (100g): sometimes it dropped to 2oz (50g)
Milk	3 pints (1.8 litres): sometimes it dropped to 2 pints (1.2 litres)
Sugar	8oz (225g)
Preserves	1lb (450g) every other month
Tea	2oz (50g) no registration was required
Eggs	1 egg a week: at times dropped to one every two weeks
Dried egg	1 packet every four weeks
Sweets	12oz (350g) every four weeks

There was also a monthly points system; sixteen points allowed you to buy a can of fish or meat, 2lb (900g) of dried fruit or 8lb (3.6kg) of split peas.

All householders had to register with shopkeepers. Registration put an enormous extra burden upon retailers. Many householders who found they had

to fill in numerous spaces (in block letters) with signatures and retailer's name and a reference leaf at the end of the book lost patience and passed the job to the retailer. Sitting up half the night filling in ration books for customers seems to have become a common experience. When the pages were complete, the appropriate counterfoils from the books had to be detached and forwarded to the Food Executive Officer. Leighton Buzzard Industrial Co-operative Society still gave dividends on all purchases.

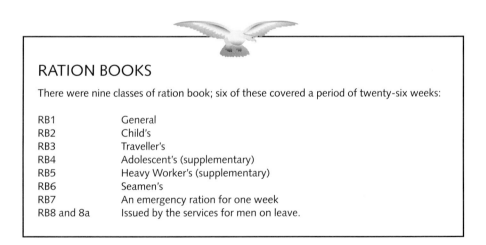

RATION BOOKS

There were nine classes of ration book; six of these covered a period of twenty-six weeks:

RB1	General
RB2	Child's
RB3	Traveller's
RB4	Adolescent's (supplementary)
RB5	Heavy Worker's (supplementary)
RB6	Seamen's
RB7	An emergency ration for one week
RB8 and 8a	Issued by the services for men on leave.

The general ration books for most adults were buff in colour. Pregnant women, nursing mothers and children under 5 were issued with green ration books and had first choice of fruit, a daily pint of milk and a double supply of eggs. Children aged 5 to 16 had blue ration books entitling them to fruit, the full meat ration and half a pint of milk a day. Vegetarians could exchange their meat coupons for other foods. Evacuees were given their own allocation and the ration given to each child before being taken to their billets was more than generous: indeed, for many tots a 1lb tin of beef, several tins of milk and a bag of biscuits seemed more than a trifle absurd. The local authority's intentions were good, and erred, if anything, on the right side.

A local issue arose when the Wing and Linslade Joint Food Committee took exception to the extra rations of bacon and butter for RAF girls. The Food Executive Officer, Mr M.C. Clifford, wrote to the Minister of Food regarding the allowances of 14oz (375g) of bacon, 7oz (175g) of butter and 21oz (565g) of sugar to the female typists at RAF stations billeted in private households. The minister replied that the Service ration book was issued to full-time members of the armed forces and the WAAF was included in this category: it was a matter for the employer and its personnel to consider further if necessary. If WAAF members were billeted in private households, arrangements could be made between the landladies and the girls as to the rations they required.

The Ministry of Food then decided to ration jam, marmalade, syrup and treacle and announced that consumers would be entitled to 8oz (225g) of preserves each month. This was received with mixed feelings. Housewives who had been unable to buy preserves for a month thought there should be a more equal distribution. The local press in Buckinghamshire and Bedfordshire pointed out that a large

The ration book issued on 7 July 1941 to Edith Skevington of 46 George Street, Leighton Buzzard. Inside are the names and addresses of her 'present retailers' for meat, cooking fats, butter and margarine, bacon and sugar. She bought her meat from Grace's in the High Street and everything else from Willards, the town grocers at 47 High Street.

Edith Skevington's identity card. These had to be carried by everyone and produced 'on demand' by a police officer in uniform or an on-duty member of HM Armed Forces in uniform. Any breach was punishable by a fine and imprisonment.

An extra ration book for additional food beyond the staples, like imported goods, that might become available. Mrs Skevington ran the Willson's shoe shop at 39 High Street. During the war Edith Skevington had three assistants, all men over 70, whom she called the three musketeers.

proportion of housewives made their own jam and reckoned that if extra sugar was allowed any shortage would be readily overcome. Shortages, as at a dance organised by the Scouts Committee in the Leighton Buzzard Gymnasium on 1 June 1940, meant patrons had to bring their own sugar for refreshments.

Retailers reported that many dodges were used to avoid surrendering coupons and they had to keep a constant watch to ensure the right number were given up. One farmer surrendered the correct clothing coupons for a pair of boots and then demanded to be excused payment as he claimed the coupons had a monetary value.

At the height of the U-boat blockade, the meat ration was reduced from 1s 2d per head per week to 1s (5p). Meat supplies in Leighton Buzzard and Linslade were smaller than for many weeks past and consumers unable to obtain the full ration of fresh meat had to make do with corned beef. This shortage was apparently due to the Whitsuntide holiday, when deliveries of meat on two days fell far below requirements. Leighton Buzzard butchers managed to supply most of their mid-week customers, but to meet demand when supplies improved, Linslade butchers made deliveries in the evening, and for the first time in many weeks opened their shops on Thursdays, commonly a half-day closing. The Ministry announced that offal, which included hearts, kidneys and liver but not ox skirt, would be de-rationed throughout the summer months.

The Ministry of Food ordered milk dealers to reserve a day's supply so it could be used in the production of dried milk and cheese. Despite the Ministry of Food's announcement that the cut should be spread evenly over the whole week, the Milk Marketing Board favoured non-delivery on one day a week and housewives were informed that no milk would be delivered on Thursdays in future and were warned to keep enough milk for use on that day and Friday morning. Some households waited patiently until midday for the milkman to call, and then suddenly realized there would be no delivery. These women were to be seen hurrying to the nearest dairy with a jug. There was some doubt as to whether housewives might go to farms and dairies on Thursdays to buy milk.

Rationing of meat led to families trying to supplement their meat allowance with fish one or two days a week. However, supplies of fish were meagre. For a week in December 1942 there was a particular problem: the total supply of fish for 15,000–20,000 people was only 29 stone (185kg). The whole of this arrived in one delivery at Messrs Bardell Bros and had sold out in little over an hour. Scores of housewives had to go home empty-handed after queuing for hours. Venison and rabbit were not rationed but these were difficult to obtain from shops.

In June 1943 a county-wide appeal led to the collection of 3cwts (150kg) of stinging nettles for a 'secret wartime purpose'; nearly 7 tons of rose hips for making rose hip syrup to provide Vitamin C for children in place of orange juice; and over 27 tons of horse chestnuts as a new source of glucose and, as by-products, fire foam used in extinguishers and a 'substance used in making munitions', according to the *Leighton Buzzard Observer* in June 1943.

It was nine years after the end of the war before rationing finally ended. An attempt to de-ration sweets in 1949 failed as demand far outstripped supply. By 4 July 1954 the last items were de-rationed – bacon and ham.

Petrol Rationing

Petrol rationing began on 22 September 1939. The government urged people to cut down on their journeys as much as possible and posters displayed the saying 'Is your journey really necessary?' Each motorist subsequently received a monthly allowance of between 4 and 10 gallons (18–45 litres). Petrol required for civil defence purposes was coloured with a special dye to distinguish it from the civilian supply, and adequate provision was also made to guard petrol supplies since it became a highly valued black market commodity.

A meeting was called at the Swan Hotel on 21 September to hear a statement on petrol rationing and the pooling of resources. At first, local garages had little difficulty in meeting customers' demands, due to a rush to buy petrol during the earlier part of the week when customers had filled up to capacity. Some customers, not content with filling their tanks to capacity, tried to squeeze in an extra gallon after the gauge said 'full'. In most cases they found the tank would take that extra gallon but shortly after it went through the overflow into the road.

In cases of real necessity applications could be made to the Petroleum Officer for additional supplies. A form for this was obtainable from the Post Office that issued the ration book, and full details of the purpose for which the fuel was required had to be given.

Meanwhile, the effect on garages and the motor industry was little short of disastrous. It is thought that tens of thousands of motorists laid up their vehicles; and others tried to run them on paraffin or even coal gas – both expedients used in the last war. The initial rationing lasted two months. The coupons were valid for fourteen days after issue; beyond this they were useless.

Worst hit were commercial vehicles and local traders. All commercial licensees were asked to register voluntarily before rationing came into force and were told that in the event of war they would be organised into appropriate groups for rationing petrol. Anyone outside the groups would get no allowance.

Through the Leighton Buzzard and Linslade Chamber of Trade, Mr Stanley Garner Cook, of Leighton Buzzard Motors in the High Street, was appointed organiser of the group owning vehicles under 2.5 tons; Mr Arthur Christopher Biggs of 1 Grovebury Road for those over 2.5 tons; and Mr E.C. Turney of

Stanbridge Road, for brick hauliers, etc. Mr Cook had more than 100 owners on his list, which covered neighbouring villages over a wide area as far as Houghton Regis.

One lorry driver remarked that the petrol he would get would run his lorry for only two days a week and then he would have to lay it up:

> Unless I can get Government work and allowances of petrol to keep the lorry running during the rest of the week, I cannot possibly pay the licence duties and driver's wages. I have bought a horse and cart and I shall restrict my work to journeys in and about the town.

Traders were also asked to devise a pooling system whereby one baker or butcher could cover one zone of the district. But this was not popular as each trader thought the other wouldn't provide the same high-class service to their customer. Also all traders' vehicles were full of their own goods on the outward journey so it would be unlikely that they could add anyone else's goods to their vehicle. However, one butcher said that he was only able to complete his rounds on his petrol allowance if he cut out two deliveries daily.

Milk deliveries were seriously affected, being cut down to once a day; and bakers were instructed by the government not to bake or deliver bread on one day each week. Thursday was decided upon in Leighton Buzzard. There is no doubt that pooling deliveries resulted in enormous savings for the traders. Supplementary rations of petrol were not given for retail deliveries as the government believed they were unnecessary in times of war.

The petrol shortage saw the trade in old carts and trolleys increase. Everybody who could cut out short trips for petrol-driven cars did so and the horse-drawn cart temporarily came into its own again. Carts that had been sold for £1 before the war changed hands at anything between £5 and £10. Useful box carts made absurd prices.

The restrictions also started a run on cycles old and new. People who used to get out the car for every shopping trip started to make their rounds on bicycles. Townspeople never before seen anywhere but 'at the wheel' were now to be seen 'on wheels'.

The Retail Trade and Public Utilities

The fears aroused by the slump in retail trade during the three weeks before the war started came to nothing. The coming of thousands of London children and mothers to Leighton Buzzard, Linslade and surrounding villages gave an unanticipated boost to local trade. This was merely a repetition of history:

Leighton Buzzard was reputedly never more prosperous than during the months Lord Kitcheners Army was billeted in the town in the autumn of 1914. Some retailers saw an opportunity in offering items such as torches for dark nights and woollen underwear, heavy woollen socks, wool felt hats, and 'real good quality Raincoats and Mackintoshes' for Special Constables and LDVs on night duty.

The Home Office Order made under the Defence of the Realm Regulations, requiring shopkeepers to close at 6 p.m. in the week and at 7.30 p.m. on Saturdays, came into force on Monday 30 October 1939. The only exceptions were confectioners and newspaper shops. The earlier closing times were received with mixed feelings. Businesses were willing to close at 6 p.m., but they wanted extra shopping time on Fridays and Saturdays. Most workers were paid on Friday afternoons and as most did not reach home until 5 p.m. housewives could not go shopping if the shops closed at 6 p.m. Tobacconists and confectioners were exempt, but butchers and grocers had a grievance. The position was none too pleasant for the general storekeeper who could keep open until 8 p.m. for the sale of cigarettes and sweets but was debarred from selling foods.

Local traders reported 'runs' on all kinds of goods, particularly articles of ladies' attire that might get dearer. Some ladies must have stocked up on stockings for a whole year.

Leighton Buzzard Chamber of Trade Council applied to the Bedfordshire and Buckinghamshire County Councils, the local authorities in these matters, for a general extension in Leighton Buzzard and Linslade to 7 p.m. on Fridays for all trades and for a special extension for hairdressers until 7 p.m. Monday to Friday and 8 p.m. on Saturday.

The abolition of cheap return fares to London was also benefiting local shopkeepers. Ladies who had been in the habit of shopping 'in town' had had to give it up and shop locally.

The presence of so many evacuees and other temporary residents made heavy demands on Leighton Buzzard water and gas supplies. The Leighton Buzzard Gas Company was called upon to supply an extra 10 per cent of gas and the demand for cookers was unprecedented. Water consumption also increased considerably, both in Leighton and Linslade. There were appeals to use water wisely because of shortages. The Leighton pumps managed to produce an additional 25,000 gallons (114,000 litres) a day; and the increased call in Linslade was also large. Householders were requested by Leighton Buzzard Urban District Council to avoid wastage. It was suggested that the washing down of shop pavements might be done once or twice a week instead of daily.

Many householders in the district had a bucket of good Leighton Buzzard sand 'against emergencies', mainly to put out incendiary bombs. Large shallow trays of sand stood outside a Leighton Buzzard garage. It was reported that 'Never before had the Leighton Buzzard sand trade heard such a national cry for "sand, sand and more sand".' Official instructions were issued on the best way of protecting sandbag defences against the rigours of winter. The precautions suggested spraying the bags with tar, creosote and other waterproofing liquids, and regular disinfection against vermin and insects. It was reckoned that the 'perfect' sandbag wall should be from 6–8ft (1.8–2.4m) high, 4ft wide at the base, and 2ft wide at the apex. Bags were laid in alternate rows of stretchers and headers and were to be only three-quarters full so they could be beaten into a rectangular shape and placed with the seam upwards.

When restricted lighting came into force it was noted that between 8.30 and 10 p.m. on a Saturday evening the main street, some of the side streets, and most of the businesses and private residences had complied with the regulation and were in almost total darkness. All roadside kerbs at ends or on corners of main roads had been painted white; and white lines placed down the centre of main roads to guide traffic during the blackout. The lines proved advantageous in Leighton Buzzard Market Square where three streams of traffic converged. The subdued lighting at the Cross was quite visible to road traffic, but motorists were less careful when parking and some cars were left in the street with headlights full on. In Lake Street a middle-aged lady asked to be escorted across the road, saying 'I have been standing here for five minutes or more and am afraid to cross because it is so black.'

A Beaudesert resident reading an evening paper at 11.30 p.m. was roused by vigorous knocking on his front door and warned by two special constables that his lights were visible outside. His family had gone to bed earlier and left the hall and landing light burning. Older people found the nightly blackout much less inconvenient than the younger folk. They had learned the art of 'feeling their way about with their feet' in the last war.

Not for many years had there been such demand for hives and bee-keeping accessories. The shortage of sugar tempted many people who had given up bee-keeping to overhaul and repaint old hives, make up new frames and try once again to get what the authorities described 'as really the best and most natural sweetening agent either for the table or for cooking'. Even old-fashioned straw skeps, which were regarded as wasteful and unclean, started appearing about the countryside. A well-known Leighton Buzzard stockist of bee-keeping accessories said that the season was the busiest he could remember; he could not get deliveries of new hives fast enough.

A meeting was held at Leighton Buzzard Occupational Centre to discuss the formation of a Leighton Buzzard and Linslade Allotments and Garden Society to stir up enthusiasm for digging for victory. It was regarded as essential that every foot of ground be cultivated, thus saving shipping for the transport of war materials. The forming of such societies enabled members to obtain seeds slightly more cheaply and as trading was done locally, it helped business people.

The co-operation of local authorities was essential, because they had the power to spend money on allotment amenities and requisition land for cultivation. Members of the County Education Horticultural staff could be called upon to give advice, and the cost of running demonstrations could be recovered from the County Educational Committee. There were no financial worries as local authorities were empowered to spend money on this 'dig for victory' campaign.

There were many examples of local contributions. Mr K.E. Clarke, a member of the William Ellis School staff, was so moved by the campaign that he obtained a plot of land from Mr Frank Cecil Meeking to be dug by several of the boys and seeds planted. Although most unscientific in their methods, the boys' labours were rewarded as the seeds came up. In addition, Mr Frederick Scott Parrott offered a field in Garden Hedge for allotment cultivation. We know that there were fruit trees nearby: hence the present-day names of Plum Tree Lane and Pear Tree Lane.

Amid the influx of hundreds of uniformed men and women billeted in the town, plus the evacuees, the town adapted to its new circumstances and there was a tendency to bend the rules.

One food item that was outside the reach of rationing was rabbit. From mid-1942 Leighton Buzzard Fur and Feather Society noted a growth in rabbit keeping and the membership of the club trebled in six months. Members agreed to send half of their progeny of does to approved butchers and the young ones for breeding. Beaudesert Senior Boys School ran a rabbit club and it was noted that the healthy rabbits would 'do credit to an adults' club'. Breeds of rabbits at shows included Dutch, English, Belgian, Any Other Variety (AOV) Fancy, Havana, Chinchilla, Giganta, Any Other Colour (AOC) Rex and Havana Rex; in 1944 over 200 entries were received. Many of the less successful entrants were afterwards eaten.

Amateur gardeners were considered by the government to be vital to the war effort to supplement dwindling supplies of imported food. The people of Victoria, Australia, sent 3lb (1.4kg) of onion seed to the local Allotment and Gardens Society as a gift to the home gardeners of the 'Mother country' and these were distributed to members. The gardeners were supposed to grow only vegetables but missed the joy of seeing spring flowers. It was decided that although heating was only permitted in greenhouses for raising vegetable seedlings, pots of spring bulbs could be 'stored', as stocks of these flowers would be needed after the war.

People risking their lives going on bombing raids or doing vital war work decoding signals seemed to have a healthy disrespect for the small rules of everyday life. One of the most common charges brought by the police was cycling without lights. In July 1943 Linslade Police Court fined eleven cyclists in one week – all members of the armed forces. By the following month the court said the fines would increase from 10s to £1 for the offence, but it appeared to have little effect on the numbers being charged.

Furthermore, boys with time on their hands continued to get up to mischief. In 1943 the local paper reported complaints from readers involving young boys using the blinds in front of shops as gym equipment. 'Blind swinging' caused serious annoyance to traders, as 'small boys with overdeveloped muscles and underdeveloped brains caused the supporting irons of the blinds to bend and in one case break away from its supports'. Another complaint was of 'roof watching'. 'Gangs of children were climbing the firewatchers' ladders and prowling around the church roof. Some were also carving on the west porch – one of which was "Bill Jones loves Mary Smith". A spokesman from the Church said that "anyone caught carving anything similar should be charged the statutory fee for wedding banns".'

However, boys also showed enterprise in keeping their pets alive. Ludovic McRae, one of the William Ellis boys sent to the town to escape London bombing, recalls feeding the family cat on gudgeon caught in the Grand Union Canal at the Martins, a Chinese-style mansion in Linslade. The boys used the boathouse at the mansion as a bathing hut from which they dived into the canal. The boys were also pressed into service as air-raid wardens and encouraged to keep out of trouble by using the Forster Institute in in Waterloo Road, Linslade, in the evenings for a stamp club, darts matches, chess and other activities.

17

SNAPSHOTS FROM THE HOME LEAGUE

Snapshots from Home League, organised under the auspices of the Young Men's Christian Association (YMCA), was a scheme originally devised in 1915 to organise local amateur photographers to take snapshots of the families of men serving at home and overseas and to forward them on to them. The scheme was revived during the Second World War, but was then also extended to include the families of evacuated children and men on active service in the UK. Leighton Buzzard became one of the centres for the snapshots programme. The Admiralty, War Office and Air Force actively collaborated with the YMCA in the distribution of the Request Forms among men overseas.

As voluntary workers for the League, the enrolling photographers had to agree to take snapshots of families in their own locality and to provide prints free of cost, as well as forward them on to the servicemen with a greeting from the families. If photographers could no longer participate in the scheme – for instance, if they were about to be called up themselves – they were urged to try and find another amateur photographer and 'enthuse him with the greatness of the job and get him to send his name up to HQ and register as a member'.

A local volunteer for the Snapshots scheme was local man Hubert Griffin. He ran an electrical shop in Leighton Buzzard High Street. Always known as Bert, he was well known and popular locally. He had volunteered in the early days of the war for service in both the navy and the air force, but was by then too old, and in any case his work as an electrician was considered to be of national importance and was therefore classified as a reserved occupation. Still anxious to play his part and to help in the war effort, Bert became involved in Civil Defence work, which took up much of his spare time. He also had a great interest in photography and when his attention was caught by an advertisement in *Amateur Photographer* magazine calling for volunteers to take part in the Snapshots from Home scheme for the YMCA, he quickly enrolled. Film and printing paper were made available for the purpose, but all other costs had to be borne by the photographer.

The 'Snapshots from Home League' logo used by the YMCA.

Servicemen would complete a form with the name and address of their family, which was sent to the London HQ, which sent the request to the nearest amateur photographer enrolled. Bert received requests from Luton, Dunstable, Leighton Buzzard, Stewkley and many local villages. He had to write to the family informing them that their relative had requested photographs of them and saying that he hoped to call on a certain day and could they be prepared. He could not use petrol so had to depend on buses. He particularly remembered making his way to Luton on a very hot day, finding no one at home when he got there. This was not usual, however, and sometimes several families got together to make the whole job worthwhile. He received many letters from sons and husbands, some of whom had not been able to see their new babies, and one from a mother whose son had been killed in action. She had received his personal effects, including the photographs he had received, and was most grateful. In his own words, Bert found the whole scheme 'most uplifting in those terrible days'.

One of the letters Bert received was from Walter Randall, whose home address was at that time in Bassett Road, Leighton Buzzard. Walter wrote of his great pleasure at receiving the photos, 'they couldn't have come at a better time either, for I got them last night, along with a Christmas card and Greetings card from the wife herself, altogether making a jolly mail and brightening up my Saturday night.' Mrs Randall was obviously very pleased with the photos as she wrote to Bert requesting further copies.

Mrs Marion Dollimore of North Street, Leighton Buzzard, wrote to thank Bert for photographing her baby daughter, saying 'I know John will be very, very pleased with the photo, as this is really the only way he has of getting to know his daughter.' She wrote that the offer to take photos came 'at a moment when I was feeling very down about things and you've no idea how that cheered me up, and I can assure you it will have an even greater effect on John'. John also wrote thanking Bert for the three photos of his wife and child: 'They came through without a scratch and I am very pleased with them. One cannot realise how much comfort one gets out of such photos, they help to overcome the loneliness that often creeps into our hearts.'

Lance Corporal J.H. Scott, whose home address was in Tilsworth, wrote from Malta on 14 October 1942 thanking Bert for the photos:

> They are very good and were more especially appreciated by myself in that they arrived on the eve of my baby daughter's first birthday. As I have never seen my baby you can imagine how I look forward to any photos of her and you will see that the ones you sent were indeed a source of great pleasure to me.

A request for photographs from a Luton man, Corporal A. Tilling of the 9th Royal Fusiliers Regimental Police, was returned because he had been reported missing; but Bert subsequently heard from Mrs Tilling that he had been captured and was a POW, and 'in good health'.

Eventually, in July 1946 it was decided that the Snapshots from Home League had served its purpose and was to be closed down. Bert received letters from Sir Henry McMahon, then President of the YMCA, and Betty Brooks, Secretary of the League, thanking all the photographers for their services. Sir Henry commented: 'Through the vast number of little pictures which have been taken and sent out by the League, great joy and comfort was given to the men in battle areas who had, for so long, been separated from their families at home. It has been a great and unselfish service'. Betty Brooks wrote: 'It may interest members to know that we have sent out tens of thousands of snapshots since the League was inaugurated.'

Although Bert Griffin was not able to serve his country in the way for which he had volunteered, he and the other amateur photographers made an immense contribution by bringing a ray of light into the dark lives of serving men who were living with great danger and were separated from their loved ones by great distance.

18

THE YOUNGEST SCHOOLCHILDREN AND THE WAR

On 13 July 1939, almost eight weeks before war was formerly declared, a letter was sent to Heath and Reach School from Mr H.E. Baines, Director of Education from Shire Hall, Bedford. Headed 'Emergency Measures in Schools' it bears over four pages of instructions on the 'Education of Evacuated School Children' and 'Air Raid Precautions in Schools'.

The letter refers to the necessity of practising for air raids at once:

> Every effort must be made to avoid complacency on the one hand, and panic on the other. The county anticipates there will be between seven and ten minutes warning of an air raid. The immediate plan is to disperse the children to their homes, but if they live more than 10 minutes away to place them as 'temporary guests' in a nearby home until the air raid is over. Gas masks should be issued to everyone and rehearsals made on how to use them. One of the principal reasons for this is to make small children used to the sight of people wearing them.

As far as evacuees are concerned, the letter says that if necessary the school will have to work in shifts, the local children going to school in the morning and evacuees in the afternoons, since there was not enough room for everyone at once. If there were a shortage of furniture teachers should borrow locally or simply get the children to sit on games mats:

> In the early days of the evacuation, difficulties will undoubtedly arise, as it is quite impossible to foresee exactly what will happen. It is possible also that the children and teachers coming into Bedfordshire will be quite unfamiliar with country life. I have no doubt that these difficulties, whether of a personal or material nature, will be met in a helpful and tactful spirit.

Meanwhile, the log at St Barnabas Infants School, Linslade, at that time still held in what is now the Community Hall, described the school attempting to carry on as normal as the war unfolded. There was as much about the children's coughs and colds, chickenpox, mumps, scarlet fever and other epidemics as there was about the 'emergency'.

The first entry that is out of the ordinary comes as early as 26 June 1939, when the log records 'one refugee admitted'. There are no details.

In September 1939 the school holidays had been extended for a fortnight because of the outbreak of war. The new term opened on 19 September; and fifteen new Linslade pupils plus fifteen evacuees were admitted.

The next reference to the war comes on 5 March 1940, when the school's two classes had to be amalgamated owing to the shortage of coke for the fires. Neither classroom could be used because of the cold and everyone was taught 'in the hall'.

The Whitsuntide holiday was cancelled because of the war and ARP were practiced. It was noted that there were eighty-seven children on the roll attending two classes. On 24 July that year the children were taken into the St Barnabas church vestry for fifteen minutes as a practice in case of air raids.

The log recorded that there were fifty-eight children on the school roll on 13 September, but more evacuees began to arrive. Nineteen new ones were admitted on 18 September, and another fourteen the following day. A final intake of two children on 19 September made thirty-five evacuees in all. A new evacuee teacher also arrived – a Miss Coles from Erith in Kent. She took a new class in the hall. By 7 October there were seventy-three Linslade children including some 'unofficial evacuees', plus thirty-seven official ones. Over the next few days, some lessons were interrupted by air raid warnings, the children having to shelter in the vestry until the 'all clear' sounded. Because of unexplained 'war conditions', school start was put back until 9.30 a.m. during the winter months.

The school log continues to record the arrival of refugees, disruptions due to air raids, but of equal importance is the measles epidemic of winter 1941. This was followed by a case of diphtheria and then chickenpox. A doctor attended and inoculated 'about 75 children' against diphtheria. Ten who missed inoculation because they were absent for other illnesses were taken to see the doctor later.

Through the summer, school continued with frequent reminders to the children how to use gas masks, and drills in case of gas attack. The school had to remain open during holidays and teachers were on duty 'to supervise' evacuee children. By September 1941 the school was in the grip of a whooping cough epidemic, which lasted until the end of November, causing continuous low attendance. By then fear of invasion meant that home defence committees were organising rest and emergency feeding centres. One of them was to be at St Barnabas church.

Christmas was celebrated as usual, each child presented with a gift from the tree. As far as possible, life continued as normal, although school remained open during the holidays for 'games, walks etc under the supervision of a teacher'.

On 16 July 1942 'Class 1 gave a short display of old English games on the vicarage lawn at the Garden Party held there.' As time passes there are fewer mentions of the war, although gas masks were still tested. There were only three evacuees still at the infant school at the start of the autumn term in 1943, most having moved up to the juniors. The school roll had gone down to sixty-six, split into two classes.

During November the first case of mumps hit the school and in December it became an epidemic. Four cases were reported on 2 December, two the next day, four more on 6 December and another thirteen the following week. By this point the wartime issue of milk for potentially undernourished children had started and children collected their ration from school even during the holidays. Distribution of cod liver oil and orange juice for all children began on 25 January 1944.

Air raid warnings increased in 1944, probably because of suspected V1 and V2 attacks. Gas masks were still being refurbished and drills taken seriously.

In the winter of 1944–45 measles, whooping cough and colds reduced attendance to twenty-five out of a school roll of fifty-six, of which six were still classed as evacuees.

On 8 May 1945: 'VE Day. As today and tomorrow have been proclaimed national holidays, school will be closed and reopen on Thursday May 10.' On May 1945: 'Attendance very poor especially in Class II owing to epidemic of measles.'

With the end of the war in sight, the school log reported that stores and equipment were removed from St Barnabas church, the danger of invasion having long passed. The end of the war came with VJ Day on 15 August 1945, but the victory over Japan was not recorded in the school log since it was the middle of the school holidays. The final reference to the war comes on 9 September 1946: 'The two evacuee children – Irene and Kenneth West – now transferred to ordinary admission register and to be counted as Linslade children.'

19

GERMAN AND ITALIAN PRISONERS OF WAR IN LEIGHTON BUZZARD

Bedfordshire was one of only twelve counties that responded to a government suggestion about using Italian POW labour. On 26 March 1941, the Bedfordshire War Agricultural Committee indicated that use could be made of 500 men in any part of the county that was not a prohibited military area. On 2 January 1942 the Ministry of Agriculture and Fisheries announced that ten hostels were to be opened for Italian POWs in the county, including one at Leighton Buzzard. Under a trial scheme POWs would live under guard in the hostels at night but would be released into a farmer's custody during the day. The Leighton Buzzard hostel mentioned in the records was either the hutment in Hockliffe Road that was taken over by the WLA or a camp of timber-built huts at the factory of the Marley Tile Company in Stanbridge Road. Both were occupied at some point, but not for long. Some fifty Italian POWs arrived on 27 January 1942 but the Air Ministry and other authorities objected to POWs working on farms near the secret RAF communication centre and the prisoners were transferred on 19 February to a hostel at Ampthill.

In contrast, many Italian POWs who worked on farms at Little Horwood in Buckinghamshire attended the funeral in February 1943 of Mr George Gee, the owner of Manor Farm at Little Horwood. Despite the secrecy of the special communications work at the farm, it was apparently not unknown for POWs to wander in from the fields and peer through the windows.

POW camps in Bedfordshire, as in other counties, were open for inspection by the International Committee of the Red Cross. The majority of POWs were sent out to work, including draining, ditching and harvesting. Denis Argent records in a diary entry in November 1941 that some Italian POWs worked alongside the United States Navy Nursing Corps at Stockgrove Park, in a spirit of goodwill that would not have been shown if the POWs had been German. He describes how one Italian, with graphic gestures, said 'The British come – the Italians run. The Germans come – the British run.' This caused laughter all round and was much repeated at the billet in the evening, when Argent was telling the rest of his section that he had been working alongside some Italian prisoners. (This was a year before the decisive battle at El Alamein in North Africa that put the Germans in full retreat for the first time.)

A small number of Italian POWs were retained to undertake the running of their camp. Initially they received the same weekly rations as British soldiers: 42oz (1.2kg) of meat, 8oz (225g) of bacon, 5.5lb (2.5kg) of bread and 10.5oz (300g) of margarine, as well as vegetables, cheese, cake, jam and tea. This regime was later amended to suit Italian tastes and the rations were increased slightly in June 1945.

Generally, the Italian POWs caused little trouble, although there was a Parliamentary storm when a half-page picture in *Life* magazine on 22 September 1941 showed four good-looking Italians walking arm-in-arm with four English girls. The photo was probably intended to show that Italians captured and taken to Britain as prisoners lived very happily; and that Britain and Italy were traditionally friendly.

The one major exception was Antonio Amedeo, who escaped with a rifle from a working party in Pertenhall, Bedfordshire, on 9 July 1943 after killing a guard, Private Charles Hands of the Pioneer Corps, with a hedging hook, reportedly because he was not allowed to talk to a local woman. He was traced and shot dead by Bernard Sidney Shelton and his 18-year-old son John Michael Shelton of the Home Guard. John was awarded the British Empire Medal on 16 November 1943 'in recognition of gallant conduct in carrying out hazardous work in a very brave manner'. This is one of the few times in the whole war when a member of the Home Guard fired a rifle in anger.

There was also a POW camp in a field south of the church at Billington and a camp for 200 German and Italian POWs, guarded by the regular army, on Soulbury Road, to the south-west of Bluebell Wood. The POWs were assigned to local farmers to help with food production, for which the farmers had to pay the authorities, and also undertook tasks such as straightening the course of the River Ouzel. An aerial photograph taken by the RAF in 1946 shows that the camp on Soulbury Road consisted of a double row of huts. These remained there until the 1950s.

As the fighting finished in 1945 some German POWs were moved into Nissen huts formerly occupied by the Czech soldiers at Aston Abbotts and a tented POW camp was created along the drive to Norduck Farm. Over 250 prisoners were kept there. Some worked on local farms until they were repatriated over a two-year period. As Italy became a co-belligerent of the Allies in 1943, it is not clear why Italian POWs were not repatriated then. The last Italian was repatriated as late as the spring of 1948.

Linslade Wood with the huts of the POW camp next to the farm buildings. The huts were originally built as billets for the RAF. At the opposite end of the wood on the Stoke Road, the Home Guard built a machine-gun post to defend the railway bridge and the town from any potential German advance from Bletchley.

In August 2007 an 81-year-old ex-POW in Leighton Buzzard, Dr Gerhard Buntrock of Dresden, revisited the town. He recalled that he had first been sent to a camp in Stanbridge Road and later to the old Workhouse in Grovebury Road, from 1946 to 1948. He had helped to build the Lee family garage, which used to stand in Regent Street, and also worked for the Horton family in Totternhoe, who ran a butchery and a farm. Another visitor was ex-POW Wilhelm Wiesman, then 80, who said he particularly remembered All Saints' church, which he used to attend every Sunday morning with Edith Varney of Pulford Road. She treated him like a son, cooked him meals and cared for him during his time in the town. He was given a tour of the town, including All Saint's church, Mrs Varney's house, Parsons Close recreation ground, and a building that had become part of Gossards where he had boarded. 'When I am 90 I will come back and do the same again,' he said.

20

CELEBRATIONS AT THE END OF THE WAR

VE Day

Victory in Europe (VE) Day was celebrated on 8 May 1945. Most shops in town closed an hour after the official announcement, some having sold bunting and flags at inflated prices, according to local reports, benefiting from the advance notice given by the government.

There were scenes of jubilation throughout Leighton and the streets were thronged with people, many with rosettes and flags. Even cattle and sheep in their

The children's party in Leopold Road, then unmade up so no hindrance to traffic. The children came from Springfield Road, Rock Lane and surrounding streets for the VE Day party on 8 May 1945.

pens ready for the market day sales were given splashes of patriotic colour. In the afternoon a heavy shower of rain interrupted a 'beautiful ankle' competition in Bell Close. In the evening unshielded windows showed lights and a flood-lit Church Square was the focus for late-night celebrations, pubs remaining open an extra hour until 11 p.m. An effigy of Hitler was burnt in Eggington.

The following day many streets held parties for the children with tea and games enjoyed by all. Some forty lucky children at the Ashwell Street party were presented with half a crown (2s 6d) and a bar of chocolate before enjoying a display of fireworks.

VJ Day

Just three months later, on 15 August 1945, the town celebrated Victory over Japan (VJ) Day and once again the town was decked with flags and bunting for a two-day holiday. Church bells rang, thanksgiving services were attended by many and the celebrations ended with music, dancing in a floodlit Market Square and fireworks. Alas, many of the pubs ran out of beer.

There was high excitement and expectation as news started to filter through from the Far East about men who had been prisoners of the Japanese since the fall of Singapore in early 1942. The postmen and telegraph boys in Leighton Buzzard were busy as individual communications to families brought news from India of sons and husbands who were en route to England and home. Letters were received by relatives in Leighton Buzzard and district from men recently released from Japanese POW camps and who hoped to be home in the near future. The public began to get some idea of the conditions of camp life and the appalling treatment of the POWs.

Wing airfield, because of its central location, was selected for the reception of returning POWs. The first wave of 819 men came on 9 April 1945 in thirty-three Dakotas; and between 9 and 30 April, 14,000 POWs were landed either at Wing or Little Horwood, the largest intake being 2,190 on 18 April. The repatriation continued into May; on the 18th no fewer than 132 Lancasters landed: the aircraft came in so fast they were landing before the one in front had reached the end of the runway. After being sprayed with DDT, the men emerged from the delousing room to be greeted by WAAFs in the main part of the hangar, which was decorated with bunting and coloured lights. George Bignall of Wingrave recalled that 'some of the returnees had tears in their eyes at the sight of the WAAFs, for it had been so long since they had seen a woman.'

For some of the younger children this was when their 'picture' daddy became a real-life father. But it was also a time for remembering all the sons, sweethearts and husbands who had given their lives for their country and had not returned.

PART THREE

WHERE THEY SERVED

Like every other place in Britain, many of the young able-bodied people of Leighton Buzzard, Linslade and the surrounding villages volunteered to fight for their country. While the town was being used as a base for a secret war waged by scientists, linguists and other specialists, the urban and rural communities were mobilising to fight the Germans, Italians and Japanese in more traditional ways.

Here are some of the stories of those who went off to fight in the various armed services. Many survived, some were imprisoned and suffered greatly as prisoners in Burma and elsewhere, and others died in the service of their country.

Men from Leighton Buzzard and district served in numerous army units throughout the world, from the initial defeats to the final outcome. The following sections give a flavour of the hell and high water they went through, all of it with a stiff upper lip. One young man, Private Cecil Henry Turney of Luton, was so anxious to get into action that he stowed away in a landing craft and hitchhiked to the battle line at Aachen, where he was found firing at the enemy. Despite deserting his unit, he was allowed to stay with the army and, indeed, he was promoted.

21

DUNKIRK

The Retreat from Dunkirk, France, 1940

In a rather understated report to the *Leighton Buzzard Observer*, an RA officer from the town described his war as involving 'some exciting experiences':

> We took part in the retreat from Dunkirk. At 8.30 one Sunday morning we were ordered to take out the breech blocks of the guns. I was driving a wagon but eight miles from the front we were ordered to stop. Then we had a march of 30 miles to Dunkirk, an experience I don't want to live through again! At one moment we were marching along Army style; the next the drone of aeroplanes was heard and we had scattered, burying our heads in ditches, hedgerows or any other place that was available. Late on the Monday we got to Dunkirk, tired, thirsty and hungry, but we had to search deserted houses before we could get anything to eat or drink. Then we, and thousands of other troops, waited on the sands for ships to take us off. Our destroyers and Air Force did wonderful work, but we were under constant bomb attacks and machine gun fire. Once the Germans came over and bombed us 17 times in a few hours. Talk about ostriches! We were smoking at one minute and the next we had our heads buried in the sand. We were helpless, but our captain was a good sort and got us off from the pier. Others waded neck deep to reach the boats.

Asked if the morale of the troops was affected, he answered an uncompromising 'No'. 'We shall beat them,' he concluded, 'but it will need the best efforts of the Navy, Air Force and Army.'

The losses incurred in the French campaign, culminating in the Dunkirk evacuation, were staggering. The RAF lost over 1,500 aircrew, the RN over 200 ships, the RA 2,500 guns and the British Army had over 68,000 casualties, including 12,000 killed. But 338,000 men rescued by 800 boats in the 'Miracle of Dunkirk' survived to fight another day.

CASSINO

Attack at Cassino, Italy, 1944

The Bedfordshire and Hertfordshire Regiment played a vital role in taking the Italian town of Cassino in May 1944, against very strong German opposition. The regiment crossed the River Rapido by ferry on 14 May. The following day they were heavily shelled and the Germans launched a fierce counter-attack, which the regiment helped to resist. The Germans fell back and it was the Allies' turn to attack.

Describing their taking of Hill 59, to the west of the River Gari, a military observer wrote in the *Leighton Buzzard Observer*:

> The Germans knew the importance of this hill, which dominated the surrounding countryside for several hundred yards, and that it would have to be taken by a famous British infantry division if the breakthrough of the Gustav line were to be a complete success. For these reasons he (the enemy) was prepared to fight hard. A platoon of the Bedfordshire and Hertfordshire regiment was chosen to lead the attack. It had been battling strenuously for three days and its numbers were depleted, but the men retained their grim determination to win through.
>
> Under Corporal Bob Lamplough, their leader, these soldiers went into the assault, completely disregarding the heavy shelling and mortaring. Lamplough led his men like a seasoned commander through the machine-gun bullets pouring almost unceasingly from the *Boche* machine-guns. Cheerfully calling his men to get in amongst the Huns, he led the close quarters fighting with absolute fearlessness, and, when only ten yards from the enemy's outer strong points, hurled hand grenades into the Spandau posts and followed through with fire from his Tommy gun. A German officer rallied nine of his men to beat off our soldiers, but Corporal Lamplough singled out the officer, and with his comrades close beside him, launched a fierce attack, against which not even Hitler's paratroopers could stand. The capture of that point laid the way open for our troops to advance through to Route 6 and capture Cassino.

The Regimental History records in its account of this action that the 2nd/Beds and Herts began crossing the River Gari on 12 May. At 6.30 p.m. on 16 May the battle began for the 'Gustav Line', a defensive line across the breadth of Italy, the key feature in the west being the massif of Monte Cassino. Company A (including Lamplough) attacked Point 50 on the right and D Company Point 58 on the left, with Company C in reserve (rather than Point 59 quoted above). Almost at once D Company was halted by very heavy fire but A Company, although suffering heavy casualties, which included Major 'Ginger' Rayner, the company commander, made good progress. Led by Captain Horace Hollick, they reached their objective by 8.40 p.m. By this time A Company had been reduced to one officer and twenty-nine other ranks and D Company to one officer and thirty-eight other ranks. Confused fighting continued throughout the night but Point 50 was held throughout 17/18 May. Several counter-attacks on the 18th were beaten off, although the position was bombed from the air and shelling at times was very heavy. The battalion had gone into action with twenty officers and 450 men and when relieved five days later had only five officers and 200 men left, a very heavy loss of 75 per cent of officers killed or wounded and 50 per cent of the overall strength killed or wounded. A, B and C Companies had to be amalgamated into a single company. But the Gustav Line had been broken, the way to the town of Cassino was clear, and the enemy was in full retreat towards Rome.

23

HMS ARK ROYAL

Billington's Youngest Crew Member on HMS *Ark Royal*, Lost Near Gibraltar, 1941

The *Leighton Buzzard Observer* of 2 October 1941 reported that Boy Telegraphist Bernard Evans, of the lost aircraft carrier HMS *Ark Royal*, had returned to Billington on thirty-three days' leave. He had celebrated his seventeenth birthday only a few days before *Ark Royal* was sunk. She was his first ship and he had been serving on her for four months, after joining the Royal Navy as a regular in 1940 and passing through the Shore Training Establishment HMS *St George* on the Isle of Man. The ship he was first posted to was sunk before he joined it, and he was left stranded on the dock at Gibraltar. Being a boy telegraphist, however, he was quickly assigned to the radio crew of HMS *Ark Royal*.

Boy telegraphist Bernard Evans, the youngest crew member of HMS *Ark Royal*, in 1941.

Describing the torpedoing of his ship, he said that at about 3.50 p.m. he was at tea when there was a loud crash. Everybody made for the flight deck and the ship took a list within five seconds. It was not realised that a torpedo had struck her until the captain gave the order 'To action stations'. When it was seen that the ship still had a heavy list, a further order to 'Abandon ship' was given. By that time an escorting destroyer had tied up aft, and those who could do so jumped onto her or shinned down ropes to her deck. Evans slid down a rope from amid-ships to a 'Carley-float' in the water, and about thirteen men climbed on and paddled to the destroyer. Between 900 and 1,000 ratings were rescued.

The destroyer spent the next four hours circling the *Ark Royal* dropping depth charges. Evans thought the attacking U-boat could not have escaped as they were only 40 miles from Gibraltar and there were many anti-submarine boats in the vicinity. About 9 p.m. a motor launch took a party of electricians and engineers back to *Ark Royal*. The others, including Evans, were landed at Gibraltar. All expected to see *Ark Royal* towed into Gibraltar next morning, but she went down 25 miles (40km) out at sea, have stayed afloat for between fourteen and fifteen hours.

Evans' parents, William and Allison, received no news of him until he returned home on leave. Bernard, nicknamed 'Lofty Sparks' as he was 6ft 4in (1.9m) tall, later served on the destroyers *Cotswold* and *Eastern* on Russian convoys. He also served on the *Illustrious* and at several shore bases. He was invalided out of the navy in late 1946. He died in 2002 and is buried in Nelson cemetery in New Zealand.

Another son, William Evans, known as 'John', was serving on *Southampton* and *Terror* when they were sunk. At the time *Ark Royal* was sunk, he was in the eastern Mediterranean and had been absent from home for two years. After the war he stayed on in the navy and was involved in mine-sweeping off the Albanian coast.

24

VENTOTENE ISLAND

A Linslade Man Describes the Capture of Ventotene Island, Italy, 1944

Lieutenant Commander John Ivester Lloyd, of the Royal Naval Volunteer Reserve (RNVR), son of Florence Ivester Lloyd of Linslade and in peacetime a writer on sporting and country subjects, described in the *Leighton Buzzard Observer* of 4 January 1944 the adventures of a flotilla of motor launches that landed paratroopers and captured Ventotene Island, some 35 miles (55km) west of Naples, before the Allies landed at Salerno:

> Six motor launches of our flotilla were in the force that attacked the island. It was a pitch black night. The force was under the command of Captain Andrews of the US Navy; and we had a US destroyer with us. On board we had Lieutenant Douglas Fairbanks Jnr of the US Navy and an American war correspondent, John Steinbeck.
>
> Captain Andrews gave the Italians a time limit in which to surrender. They did, sending up the rockets required, but no sooner had they done that than the Germans sent up red rockets. Obviously they weren't surrendering. We made our way in cautiously. We had only inches to spare in getting into the harbour, where we got alongside an E-Boat at the jetty. At that moment a German sergeant on board blew up the E-Boat, but my chief motor mechanic, W Neville of Windlesham, with great presence of mind seized a fire extinguisher and leapt on board the enemy boat and put out the fire that had started. It was a smart piece of work.
>
> We landed our force of parachutists who soon rounded up the Germans, although heavily outnumbered. We took prisoner the sergeant on the E-Boat. The Italian troops on the island were all lined up to meet us, smart in their best uniforms, except one unfortunate fellow who was there in his pyjamas and kept yelling 'Me no militaire', a plea that didn't impress the US parachutists one little bit. He was marched off with the rest, hopping as best he could over the stones.

The harbour entrance was so narrow that the senior officer would not allow his other boats to go in. In any case there was no need. As soon as the island had been captured, the motor launches went off to Salerno to take up duties off the beaches.

John Steinbeck, author of *The Grapes of Wrath* and of *Mice and Men*, served as a war correspondent for the *New York Herald Tribune* and worked with the Office of Strategic Services, the predecessor of the United States Central Intelligence Agency (CIA). Some of the writings from his wartime career were incorporated in the documentary film *Once There Was a War* (1958). Lieutenant Commander Douglas Fairbanks Jnr of the US Navy 'Beach Jumpers' was made an Honorary Knight Commander of the Order of the British Empire (KBE) in 1949.

After the action, it was found that Ventotene island had a strong German garrison of 114 men to defend a key radar station. Only forty-six American paratroopers, from the 509th Parachute Infantry Battalion, were landed. They quickly met a local exile from the Italian mainland, who then fed false information to their German commander that there was a regiment of paratroopers on the island, disembarked by a fleet of Allied ships. Shocked, the German commander demolished his own positions and weapons, and then surrendered to the weaker American force before realising his mistake. Ventotene was liberated at 3 a.m. without a shot being fired.

25

SALERNO

One Man's Experience at Salerno, Italy, 1944

The *Leighton Buzzard Observer* of 30 November 1943 reported that Anthony 'Tony' Capp of 21 Beaudesert, Leighton Buzzard, a naval cook serving in the Mediterranean, had sent his parents an account of his experiences with the fleet during the Salerno landing. He was quoted as saying: 'Salerno, seen through the eyes of the men of the minesweeper trawlers, was mine-hunting by day, submarine-chasing by night, and a 24-hour heavy bombardment from the sky.' When the Allied invasion fleet were two hours out, they received orders and the group to which he was attached was told to sweep away a minefield protecting Salerno.

At midnight on the second day they neared Salerno and preparations for action were completed. Lights were doused as hostile planes were overhead. They hove-to about a mile (1.6km) off shore. Then the doors in the many landing craft swung open and a steady stream of amphibious vehicles came out, filled with Allied troops, and sped past to the beaches. The night was broken by the rasp of machine gun fire as the Tommies forced a bridgehead. Gun flashes illuminated the landing operations and enemy shore batteries opened up from the beaches. Overhead, enemy aircraft lit the heavens with coloured flares. Bombs fell as the Oerlikons sent out lines of fire, and landing craft constantly headed towards the shore.

Tony Capp:

A field gun ashore got our range and shells fell in the waters around us. A British cruiser opened up in defence, sending death into the enemy's mountain strong points. The signal 'proceed to sweep' was received. Float, sweep kite and wire were paid out and cutters. Slowly we steamed into the Italian minefield. The German field gun was determined to stop us. Every few seconds a shell dropped near us. Rivets rattled and steel plates vibrated as the sharpshooters began to explode surface mines with rifle fire. We continued sweeping throughout the afternoon and evening.

The German field gun, though not accurate, was still being annoying. Ashore the bridgehead had been established. Aboard ship, cooks left their action stations to whip up a hearty meal of bread and cheese for the sweating gun crews. Boiling tea, tainted with the smell of cordite fumes, was eagerly swallowed by parched throats. Night brought cool relief, but no rest. Guns still barked at the *Luftwaffe* while reinforcements poured ashore. The minefield had been swept, and we steamed out into the bay to keep a continuous anti-submarine watch until all landings were complete. Over the radio we heard Admiral Sir Andrew Cunningham's message that the Navy had not lost a man during the landing operations. We smiled. We had a reason. There were no longer any mines at Salerno.

26

POWS IN JAPAN, ITALY AND GERMANY

Leighton Buzzard POWs in Japan

The fall of Singapore on 15 February 1942 involved many casualties among Bedfordshire men. The regimental history of the 5th Bedfordshire and Hertfordshire Battalion TA (55th Infantry Brigade, 18th British Division) records forty-five dead in the campaign; the battalion also suffered seventy wounded and thirty-five missing by mid-1942. In the regimental history, Major R.G.B. Thompson says that, like other units, the battalion was just frittered away in bits and pieces and 'someone ought to be slowly slaughtered for the way the show developed'. Most of the Dunstable Artillery Territorial Unit, the 419th Battery of the 148th (the Bedfordshire Yeomanry) Field Regiment RA and the 18th British Division, who had joined up en masse on the outbreak of war, were killed, wounded or captured.

Henry Ellis Isidore Phillips (later Sir Henry Phillips, CMG, CBE) was one of the officers of the 5th Battalion. He regarded the use of the 18th Infantry Division in Singapore as a tragic mistake. The division had been destined for the Near East and trained in mobile warfare against the Germans and Italians. Its men knew very little about the Japanese and jungle warfare in Malaya. The division was sent into action almost immediately after being diverted to Singapore, the chain of command was disrupted and it fought as units and sub-units mixed in with other formations in a campaign that was, in Phillips' view, already irretrievably lost due to the overwhelming superiority of the Japanese in the air.

Gilbert Inglefield, later of Eggington, served as a staff captain with the 1/5th Battalion Sherwood Foresters, attached to the 55th Infantry Brigade HQ, 18th Division. His notes for 1941–43 in the Imperial War Museum describe how he travelled to Singapore, via India, on the troopships *Orcades* and *West Point*. It then covers Japanese air raids and the bombardment of Singapore, scenes, life and morale during the fighting on the island (February 1942) at the time of the

'A' Troop of the 148th Field Regiment RA had arrived in Poona (Pune), India, late in 1941. William Merry is seated fifth from the left and among the men are Derek Gilbert, 'Charlie' Greening, Bernard Hedges, Gerald Lecomte (Lewisham), John Millburn and Dennis Tooley.

surrender, captivity – including the development of Changi POW camp – his journey to Thailand in November 1942, conditions until late 1943 in Banpong, Kanburi and Chungkai camps, and cameos of individuals met at various times. He never forgot his time as a POW and the malnutrition he suffered. When he was knighted in 1965 and became Lord Mayor of London two years later, his favoured charity during his term of office was War on Want.

On 7 July 1943 Isaac and Eleanor Millburn of 38 Ashwell Street, Leighton Buzzard, received a postcard from their son, 23-year-old Gunner John ('Jack') Sumpton Millburn of the 148th Regiment RA, saying that he was interned in Taiwan (Formosa) Camp and 'was in excellent health'.

The real story was rather different. Gunner Millburn's battery, commanded by Major William Haslam Merry from Wing, arrived at Singapore on 29 January 1942 and camped near Tek Hock village on the Tampines Road. Major Merry was killed a fortnight later, but final and official confirmation of this only reached his relatives in March 1945. From 14 to 22 February 1942, Jack Millburn was in Alexandra Hospital receiving treatment for a shrapnel wound in the hand. He was lucky to survive the atrocities committed by Japanese troops when they first overran the hospital. He and other POWs were sent to Changi prison and after a month there were taken, on 13 March, to the River Valley Camp, where they were held until October 1942. The whole camp was then cleared and the prisoners, mostly British, taken to the docks to embark on a ship – the *Dai Nichi Maru* – for Formosa and eventually to Kinkaseki mine in the north, the largest source of copper in the Japanese Empire.

At great risk to himself, Jack Millburn kept a remarkable diary of his early years in captivity. He recalled that on the journey to Formosa they were 'tossed about like hell. We all have dysentery and are absolutely filthy and they will not allow the orderlies to clean the lavatories and they keep us below decks.' While working in the mine in Formosa his leg was badly injured and many other POWs were killed when a roof fell in. He said the Japanese guards were completely indifferent to the accident. Because of the starvation diet, he thought constantly about food and wrote down in his diary detailed recipes for meals he would have liked to eat. He survived the war and returned to his parents' home address.

Brothers Derek Gilbert and Kenneth Harold Gilbert of the 148th Regiment RA also became POWs when Singapore surrendered. They were taken to Bangpong and then on a 125-mile march through the jungle during the monsoon to Camp Wampo on the Burma border. Those who fell sick were dragged into the jungle and shot. At the camp they were forced to work on the Burma railway for sixteen or seventeen hours a day. Very strangely, their mother, who lived in

Isaac and Eleanor Millburn had a personal reason to celebrate the surrender of Japan: they had just heard that their youngest son John was safe. As one of the weakest of the surviving POWs, he was taken from Formosa to recover in hospital in Manila.

Stanbridge Road in Leighton Buzzard, received a postcard from Cape Town, posted on 10 May 1942, indicating that Derek 'had escaped and was safe and well'. This was not true, but perhaps it was better than knowing what was really happening to her sons in the camp.

Despite the hard manual labour, the POWs were given very little food and were forced to eat snakes, baboons, lizards and rats, grilled on sticks of bamboo like kebabs. After ten months in appalling conditions, Kenneth Harold Gilbert died of beri beri on 1 August 1943, aged 23. He was buried in Kanchanaburi War Cemetery.

A pro forma letter from the War Office, sent to the next of kin of each former prisoner sixteen days after the official surrender of Japan.

Kenneth Harold Gilbert (left) and
Donald Horne.

British POWs with Japanese guards after
the fall of Singapore. First on the right,
kneeling, is Kenneth Harold Gilbert.

While working on the railway Derek was wounded in the knee by a bayonet
thrust from a guard, leaving him with a permanent injury. He remained in the
camp for another year and a half, before the POWs were moved to Ubon in
North Thailand to build a runway for the Japanese Air Force. They were also
made to dig an enormous pit, which they were told would be their grave in case
of any attack. At the end of the war, planes from the Canadian Air Force landed
on the runway and took the survivors to Rangoon – thirty-six men per plane. At
the beginning of the war, Derek had weighed over 14 stone (89kg); by its end he
weighed 5 stone 3lb (33kg).

   For forty-eight years Derek never spoke about his experiences as a POW. He
suffered nightmare after nightmare. Then the Ministry of Defence sent him to see
a psychiatrist at the Luton and Dunstable Hospital, who told him that he needed
to talk about what happened to him. He took this advice and the nightmares
disappeared. He even visited Japan and talked to schoolchildren there about what

俘 虜 郵 便

馬來

俘
虜
檢 收
閱 容
濟 所

PASSED
C.W. 6819

Mr & Mrs MILLBURN,

38 ASHWELL STREET,

LEIGHTON BUZZARD,
BEDFORDSHIRE,
ENGLAND.

This postcard arrived in Leighton Buzzard in April 1943, the first communication that the Millburns had received from their son since February 1942. The men were physically persuaded to communicate that they were in good health and working for pay.

IMPERIAL NIPPON ARMY

My health is USUAL

I am working for PAY

I have received letters/parcels dated I HAVE NOT HEARD FROM from YOU YET BUT HOPE TO SOON. MY BEST WISHES

Please notify TO YOU ALL. PLEASE REMEMBER ME TO RITA

of my safety AND DONT WORRY TOO MUCH. KEEP SMILING.

Take care of ALL AT HOME.

I would like to hear from ALL OF YOU.

Love JACK.

it was like to be a POW. The Japanese teachers encouraged him to tell it as it was and by the end of the talk the children were in tears.

Gunner Douglas Cresswell of George Street, Bletchley, told an equally grim story about life as a POW in Japanese hands in the *Leighton Buzzard Observer* of 28 November 1944. He was also captured when the 148th Field Regiment surrendered with the rest of the 18th Division. He and other prisoners were taken to a camp on the north-west coast of Singapore and kept working on clearing debris and bomb damage until October, when they were sent by rail to Thailand in cattle trucks each holding thirty-three men. There was no room to sit down and they stood or leaned against each other for four days until they reached the rail terminus at Bangpong, from where, like the Gilbert brothers, they were marched for several days, covering 16km a day, to another camp. Here they had to build their own huts. They were given two meals a day of a pint of rice and a piece of fish about 2in (50mm) long. Red Cross parcels were withheld as a matter of course. After a few weeks building a new railway, the men began to go down with malaria. Work continued for eighteen hours a day until in 1943 when cholera broke out among the prisoners. British medical officers had no supplies or equipment worth mentioning to combat disease and performed operations with cut-throat razors. Early in 1944 the prisoners were moved to a camp at Thamvan and used to build roads, still on two meals a day. The camp housed 7,000 men and by the time Gunner Cresswell left, the latest grave was number 2,047.

In February and March parties of men were selected for transfer to Japan and were taken back to Singapore in cattle trucks. While waiting for a ship they were made to load scrap iron in the docks for eighteen hours a day, as at Thamvan. Eventually the prisoners were marched to the docks and embarked on the *Rakyo Maria*, a 7,000-ton cargo ship. The Japanese guards tried to cram 1,300 prisoners into the hold, but as fast as the men were pushed into the hold they fainted and had to be brought out. Finally the guards allowed 400 to sleep on deck but they had to go below during the heat of the day. The ship left Singapore on 6 September and four days later was torpedoed twice. The Japanese guards left in lifeboats. When they had gone, eight British officers took charge and organised an orderly evacuation using the remaining lifeboats, supplemented by hatch covers. Gunner Cresswell was in a group of seventeen men on rafts lashed together. After five days, only seven were still alive. They were picked up by an American submarine. After a spell in hospital, Cresswell was taken to Hawaii, San Francisco and New York, where he was given the best food and the best seats at shows, and then returned home, an extremely lucky man. He was given six weeks' leave and got married shortly after his return.

John Wyatt of the East Surreys had a similar experience to Cresswell's, his ship, the *Asaka Maru*, being wrecked on rocks on the journey to Japan. Picked up by a

Japanese destroyer, he was taken to Amagasaki POW camp near Osaka, arriving in August 1944. The camp was liberated on 5 September 1945, a month after the atomic bomb had been dropped on Hiroshima. As he clambered aboard the USS *Benevolence*, Wyatt was greeted by the first white women he had seen since he left Singapore: American nurses in their clean white uniforms. They looked to him like beautiful angels and he fell in love with each and every one of them, as they made him seem so welcome. It was, he said, like arriving in the Promised Land after nearly four years in the depths of hell.

The first Leighton Buzzard man to arrive back in 'Blighty' from the Far East was Private Frederick W. Drake of Heath and Reach on Saturday 6 October 1945, followed a day later by Royston 'Roy' Sutton, an RA driver aged 27 of 10 Summer Street, who had been captured at the fall of Singapore. Roy described how he had been sent to one camp in Bangkok where the POWs had had to sit upright all day and look straight ahead. Then one day a Japanese officer had lined up the POWS and said 'Goodbye, it's over'. The POWs were dumbfounded. But when they were handed cigarettes and a bar of soap each, they knew the war was genuinely at an end.

Leslie Webster from Leighton Buzzard joined the Bedfordshire Yeomanry 148th Field Regiment RA and was also captured at Singapore. He arrived in Liverpool on 13 October 1945 after three-and-a-half years in captivity. He records in his book *Life in a G-String* that there was no band to welcome the returning ex-POWs, no cheering crowd and no banners. In fact, nobody took any notice. He ends his book with the following poem, similar to the 'Ode of Remembrance', dedicated to POWs in the Far East:

> And we that are left grow old with the years
> Remembering the heartache, the pain and the tears.
> Hoping and praying that never again
> Man will sink to such sorrow and shame.
> The price that was paid we will always remember,
> Every day, every month, not just in November.
> We will remember them!

An official calculation after the war determined that of 42,610 British POWs held by the Japanese, 10,298 died in captivity. Survival rates for officers were dramatically better than for other ranks, as the Japanese spared the officers much of the manual labour and gave them a range of minor privileges. In his book *No Mercy from the Japanese*, John Wyatt states that when he received his back pay he found a deduction had been made for the rations, mostly maggoty rice, provided during his time as a POW. One can hardly imagine a more thoughtless insult.

Leighton Buzzard POWs in Italy and Germany

MAJOR H.T. PEPPER

In contrast, the treatment of British POWs in Italy was much more in accordance with international conventions, at least in as much as the local conditions and standard of life allowed. In April 1943 Major Henry Thompson Pepper of Grovebury Road, Leighton Buzzard, was repatriated from Italy, having been severely wounded and captured in North Africa over a year before in action with the Welch Regiment. In a report in the *Leighton Buzzard Observer* for 4 May 1943, he said he had been medically treated at a German base camp, looked after well and then flown to Italy. In captivity, the staple food had been macaroni, with a very small piece of meat twice a week, and about 3.5oz (100g) of bread, plus rice, daily. So the tea, coffee and cigarettes in Red Cross parcels had been much appreciated. Concerts were arranged, instruction classes held and other forms of recreation organised. His view of Italian morale was low: 'We marched out in military fashion,' he said, 'and the Italian populace trembled to see us go by.' He added that in his opinion Italian soldiers were never in the first line and were only useful as artillerymen at a range of about 2,000yd (1,800m).

LIEUTENANT R.G.M. QUARRIE

Lieutenant Richard Gordon Manners ('Don') Quarrie of the West Yorks, of Leighton Buzzard, was taken prisoner by the Germans on Crete in 1941. He was injured but cared for and flown to Greece. Then, in the cattle trucks of a train, he was taken with other prisoners via Yugoslavia to Germany. He spent time in three POW camps and escaped six times, but was recaptured every time except the last, when he got away from a column of prisoners attacked by American planes during a forced march westwards, away from advancing Russian forces. On a previous occasion he had reached Cologne, but had been betrayed by the French. This time he was hidden by a German family for two weeks and eventually linked up with Allied forces. He then had to undergo what he called 'several days of red-tape from which we received the impression … that everyone seemed to be obsessed with tracking down the perpetrators of countless atrocities'. He arrived back home in Leighton Buzzard on 2 May 1945. He was never physically abused by the Germans but life had been tough: he had usually been cold, hungry and bored; compared to POW life under the Japanese, however, he had been treated reasonably well, according to the WW2 People's War archive.

LIEUTENANT E.T. HAMPSON

Lieutenant Eric T. Hampson, RE, son of Mr and Mrs John Hampson of 'Wavecrest', Southcourt Avenue, Linslade, was captured at Tobruk and taken to

POW Camp 49 in Italy. His letters home suggested that without Red Cross parcels the prisoners would have had a very bad time. With another POW, he escaped when Italy capitulated. The *Leighton Buzzard Observer* of 21 December 1943 records that the two men thought they had failed, when at dawn, after going for sixty hours without food or sleep through snow and ice and three rivers, they were lying in a wood in the rain and a soldier came along. They thought they were still among the Boche, but the soldier turned out to be from one of the forward British platoons. After three months walking in the mountains through German-occupied territory, begging food and shelter, they had finally reached safety. In hospital, while they were recovering from exhaustion, they found getting back to English food and tobacco 'almost unbelievable' and were delighted when a special dinner was put on for them.

PART FOUR

LEST WE FORGET

22

THE MEMORIALS

The Second World War was a conflict in which people from every walk of civilian life were sent all over the world, sometimes to places they had never heard of. Those remaining in the United Kingdom would often find themselves working in jobs that but for the war they would never have thought of doing.

The majority of the Leighton Buzzard and Linslade servicemen came home again, irrevocably altered by their experiences. Of those who were killed, some died abroad and were buried there; some died in the United Kingdom or elsewhere and were buried in these islands; some were reported 'missing, presumed dead' and are named only on war memorials.

In and around Leighton Buzzard and Linslade there are more than twenty war memorials and rolls of honour in various forms, most of them erected shortly after the Great War of 1914–18. Although it was initially intended to repatriate those who died abroad as a result of enemy action, the great number of casualties and the understandable chaos surrounding their temporary burials, almost from the first battles of that conflict, rendered this plan impossible to carry out. The bereaved had been accustomed to tending graves and a dignified alternative was needed. Consequently, in 1919 local committees convened all over Britain to decide upon the form each local memorial would take. These were obviously each designed to display the number of names to be commemorated. Nobody dreamed that within so few years there would be the need to add a second list.

The names of the casualties of the Second World War have been gathered from these memorials. Men from all of these towns and villages either attended school, worked or lived in Leighton Buzzard or Linslade at some time. Several of the names appear on more than one memorial, in their places of worship, their schools and their workplaces. A few local men are also commemorated in Bletchley, Dunstable and Luton.

Some of the men commemorated were from long-standing local families; some were from elsewhere but had married local women; some had been evacuated here; some had come here either alone or with their families to find work; some had been stationed or billeted here. A few of those buried in Leighton Buzzard and

Linslade are not named on any of our memorials, but have their names inscribed on memorials elsewhere in the British Isles, in their own towns or villages of origin. Perhaps not altogether surprisingly, there are even a few casualties from this part of Bedfordshire or Buckinghamshire whose names for whatever reason do not appear on local memorials. These omissions were either by accident or by design: the town memorial erected so near the Post Office in Church Square was perhaps just too constant a reminder and thus several of our servicemen are only commemorated elsewhere, in whichever part of the world their deaths occurred. There are men mentioned on their relatives' graves at Leighton Buzzard Cemetery in Vandyke Road or at St Mary's Churchyard and cemetery in Old Linslade who are not recorded on any of the local memorials. For a few, local evidence only exists in an announcement in the *Leighton Buzzard Observer*, and further information has only been possible by referring to the Commonwealth War Graves Commission.

The Leighton Buzzard Memorial

The Leighton Buzzard War Memorial in Church Square records the name of fifty-one men killed in the Second World War (plus Major John Harold Procter,

The Leighton Buzzard War Memorial – side of the plinth facing into Church Square.

The Leighton Buzzard War Memorial – side facing All Saints' church.

The Leighton Buzzard War Memorial – side facing the Post Office.

1950). The memorial is reputed to be made of the largest single block of granite ever quarried in the British Isles. It is 25ft 3in (7.7m) high and 3ft 2in (1m) wide and deep, weighing 22 tons. The block was excavated in the Shap quarries about 1870 and taken to Messrs Farmer and Brindley's yard in Westminster, where it remained until 1920, when it was removed to Leighton. It took three days to erect the memorial. Underneath the monolith Harry Yirrell and Fletcher Brotherton placed a lead tube containing silver and copper coins of the realm. An unveiling and dedication service took place on the second anniversary of Armistice Day, on 11 November 1920. The following photographs show the inscriptions recording names of Leighton men who fell in the Second World War (lower plinth).

Linslade War Memorial

The Linslade memorial originally stood at the junction of Old Road and Wing Road, near the canal bridge. However, due to the increasing amount of traffic at this junction, it was moved in 1955 to its present position in the Memorial Gardens, Mentmore Road. The photographs on the next page show the names of twenty Linslade men who fell in the Second World War (recorded on the two lower plinths).

Linslade War Memorial in the 1930s. It was moved from the junction of Wing Road, Old Road and Soulbury Road, at the entrance to the canal bridge in Linslade, to its present site in the Memorial Gardens in Mentmore Road in 1955.

There are twenty names recorded on the Linslade War Memorial for 1939–45.

Heath and Reach War Memorial

The memorial stands in the grounds of St Leonard's church on Leighton Road, bearing five names from the Second World War.

The Heath and Reach War Memorial recording the names of John Bond, William James Hack, Frank David Puddlephatt, Frederick John Stone and John William Whiting.

23

ROLL OF HONOUR

The circumstances surrounding deaths on active service obviously rendered accurate reporting difficult: news would sometimes arrive via a dreaded telegram soon after the event but, equally, months or even years might pass before official confirmation was received. In some cases there were only ever the sparsest of details while in others full reports would eventually become available. Consequently, although the following biographies are the result of intensive research, it is impossible to know if all casualties have been identified.

One of the first casualties of war after 3 September 1939 among the servicemen with connections to Leighton Buzzard was John Morton. Six years of anger, fear and grief would follow, and many tragedies on a local and a national scale, until the last service death recorded in the *Leighton Buzzard Observer* up to 2 September 1945, that of Stanley Abbott. Between these two deaths were so very many more. Each account of the final days of the known casualties is followed by information relating to their pre-war lives, their immediate family members and the location of the grave or the memorial connected to their place of death. A guide to information about them on other memorials and rolls of honour is given in the appendices to this book.

1939 TUESDAY 19 SEPTEMBER: 917351 Gunner **John James Morton** aged 33 of the Bedfordshire Yeomanry (148th Field Regiment RA) died as he and four other soldiers were returning to the Territorial Army Drill Hall in High Street North, Dunstable after road-block duties in the town centre. In the blackout, a speeding lorry owned by Hay's Wharf Cartage Company hit them, killing John instantly and injuring the others. He was the son of John and Rose Morton of Leighton Road, Northall. St Mary's Churchyard, Edlesborough.

1939 SATURDAY 14 OCTOBER: P/SSX 19671 Able Seaman **John Brown** aged 24 of the RN was one of the 810 men lost when, on 14 October 1939, U-47, under the command of Kapitän Leutnant Günther Prien of the Kriegsmarine, torpedoed the *Royal Oak* in Scapa Bay. Scapa Flow, a body of water in the Orkney

Islands, was the site of the United Kingdom's chief naval base during both world wars. The son of Catherine Bailey (formerly Brown) of Windmill Path, Leighton Buzzard, John had served in the RN for some years before the outbreak of war. Portsmouth Naval Memorial Panel 33 Column 2.

1940 WEDNESDAY 1 MAY: 580784 Sergeant (Observer) **Peter Chris Cunningham** aged 22 of the RAF 99 Squadron died when his plane crashed during a bombing attack. In the first few months of the war, both Germany and the Allies realised the strategic importance of Norway: whoever controlled it controlled the sea routes to the North Atlantic, to the German coast, to the Arctic Ocean and to the Soviet Union. At the outbreak of war in 1939, Norway had declared neutrality but on 9 April 1940, without any formal declaration of war, Germany invaded. On Wednesday 1 May the RAF launched extensive night attacks on aerodromes at Stavanger, Fornebu and Aalborg, causing heavy damage to aerodromes and aircraft, despite strong opposition, and seven of their aircraft were lost. One of these was Vickers Wellington P9276, which had taken off from Newmarket, Suffolk, at 6 p.m. for a raid on the enemy-occupied airfield at Stavanger. Sergeant Cunningham died that night along with the other five members of his crew. The aircraft is assumed to have crashed in The Wash, since all six bodies were eventually recovered from the water on various dates. He was the son of Chris and Dorothy Cunningham of Northall and had joined the RAF upon leaving school. St Mary's Churchyard, Edlesborough.

1940 SUNDAY 26 MAY: 1505400 Gunner **Frederick William Compton** aged 21 of 58th Field Regiment RA died during Operation Dynamo, the evacuation of the British Expeditionary Force from the beaches and harbour of Dunkirk in northern France. The exact date of his death is unknown but fell between 26 May and 2 June. The 58th Field was somewhere along the Franco-Belgian border in III Corps of the General HQ Troops, 44th (Home Counties) Infantry Division of the First British Expeditionary Force on 9 May 1940, immediately before Germany invaded France and the neutral countries of Belgium, Luxemburg and Holland. He was the son of Joseph and Minnie Compton of 29 Wing Road, Linslade. Dunkirk Memorial, Département du Nord, France, Column 11.

1940 MONDAY 27 MAY: 76124 Second Lieutenant **Geoffrey Aveline Rowe** aged 34 of 1st Buckinghamshire Battalion of the Oxfordshire & Buckinghamshire Light Infantry had been a Territorial officer for three years before the outbreak of war. His regiment had been taking part in the defence of the Ypres–Commines Canal, which alone suffered more than 300 casualties before eventual evacuation. He was mentioned in dispatches in Hazebrouck, where he almost certainly died on Monday 27 May 1940. He was the son of Ernest and Annie Rowe of 'Roundway', Hockliffe Street, Leighton Buzzard. Dunkirk Memorial, Département du Nord, France, Column 93.

1940 WEDNESDAY 29 MAY: 5952111 Private **Maurice Maunders** aged 21 of 8th Battalion the Royal Warwickshire Regiment died in Belgium. The 8th formed part of the British Expeditionary Force in France and Belgium and in mid-May was among those sent to defend the Ypres–Comines Canal Zone. These troops withstood the assault of three German divisions on 27 and 28 May, playing a crucial role in holding the canal. Without their successful defence, the entire evacuation of Dunkirk would have been in danger of failing. He was the son of Arthur and Alice Maunders of High Street, Eaton Bray. Kortrijk (St Jan) Communal Cemetery Row BB Grave 5.

1940 FRIDAY 31 MAY: C/KX 103452 Stoker (2nd Class) **Norman Tearle** RN aged 20 was based at Number II barrack block of the Royal Naval Yards, Chatham, known as HMS *Pembroke*. He died during the evacuation of some of the 68,000 troops from Dunkirk on that one day. He was the son of Frederick and Deborah Tearle of Soulbury. Oostende New Communal Cemetery Plot 9 Row 3 Grave 17.

1940 SUNDAY 2 JUNE: 904032 Gunner **Alfred John Cooper** aged 18 was with 68th AA Regiment of the RA. Too young to be sent abroad, he was in a billet 'near Nottingham' at the beginning of June 1940, where some light-hearted larking about by a couple of teenagers ended in tragedy. On 2 June three 18-year-old soldiers, of whom Cooper was one, entered a barrack-room to rest. 'One of them leaned his rifle against his bed and went out of the room. Cooper picked up his own rifle, which was unloaded, and, striking a pose said "This is how I am going to charge 'em". The third lad, responding in the same spirit, picked up the rifle that had been left by the soldier who had gone out of the room and levelled it in the direction of Cooper … completely unaware that it was loaded, and the rifle went off.' A verdict of 'accidental death' was returned. This was the first of a number of local servicemen's deaths in England that year. Danger was not only to be found 'at the coalface'. He was the son of Ellen Major of Leighton Buzzard and had worked for Waterlow & Sons Ltd, Dunstable, a printing firm. Leighton Buzzard Cemetery, Vandyke Road Section O Grave 318.

1940 SUNDAY 7 JULY: C/J 101244 Able Seaman **Albert Victor Creighton** RN aged 36 was on board HMS *Mercury*, a coal-burning paddle-steamer requisitioned for use as an auxiliary mine-sweeper. *Mercury* was among the five vessels in the 11th minesweeping flotilla based on Ardrossan, on the Firth of

Clyde. As the Dunkirk evacuation began, this flotilla was ordered south to Dover to take over the Channel sweep by day and anti-invasion patrols by night. When *Mercury* was badly bombed off Portland, Dorset on 7 July, three of its gun crew died; Albert and his two shipmates are buried together in one grave at Portland Royal Naval Cemetery. Son of David and Eliza Creighton of Camberwell, he had worked for several years prior to the outbreak of war as a postman on rural deliveries in the Leighton Buzzard area. Portland Royal Naval Cemetery Church of England section Collective Plot 671.

1940 FRIDAY 30 AUGUST: George Harry Chambers aged 40, a civilian, was one of the thirty-nine employees killed during a bombing raid on Vauxhall Motor Works, Luton. Car production at Luton had been suspended to allow the firm to work on the new Churchill tank. That day a large number of aircraft dropped more than 200 bombs on the town, from the Vauxhall works as far as Caddington, killing sixty people. He was the husband of Alice Chambers of 39 Hockliffe Road, Leighton Buzzard and son of Thomas Atherstone Chambers of Masbrough, Rotherham, Yorkshire; he had held a managerial position at Vauxhall Motors. St Mary's Churchyard, Old Linslade.

1940 FRIDAY 30 AUGUST: Charles Henry King aged 55, a civilian, was killed during the same air raid on Vauxhall Motor Works, where he was working as a draughtsman. He and his family had been connected with Morgan and Company, carriage builders in Leighton Road, Linslade. He was the husband of Frances King of 51 Vandyke Road, Leighton Buzzard and son of James and Clara King latterly of 'Gayhurst', 7 Mentmore Road, Linslade. Leighton Buzzard Cemetery, Vandyke Road.

1940 SUNDAY 22 SEPTEMBER: T/140965 Driver **Victor George Sells** aged 22 of the Royal Army Service Corps (RASC) died in hospital in Amersham. He was a grandson of Henry and Anis Sells of Peddars Lane, Stanbridge and had worked in Dunstable for the butcher R.H. Bunker. St John the Baptist chapel yard, Stanbridge.

1940 WEDNESDAY 2 OCTOBER: 951796 Aircraftman 2nd Class **William George Garner** of the Royal Air Force Volunteer Reserve (RAFVR) aged 24 was 'accidentally killed on active service' near Gainsborough. The deaths of three other RAF ground crew (Corporal William Brown, Corporal Eric Godfrey Clark of 144 Squadron and Leading Aircraftman Edward 'Teddy' Hodge of 144 Squadron), who all died that same day, were registered in the Gainsborough registration district. They were almost certainly based at RAF Hemswell, used by RAF Bomber Command throughout the war, since one of them was buried locally at Harpswell. He was the husband of Bridget Garner of Lanes End, Heath and Reach and son of George and Emily Garner of Lammas Walk, Leighton Buzzard, and had been employed by wireless dealer Frank Griffin of 62 Hockliffe Street, Leighton Buzzard and by the Marston Valley Brick Company and the Bedfordshire Tile Company. Leighton Buzzard Cemetery, Vandyke Road Section O Grave 39.

1940 TUESDAY 22 OCTOBER: 1358349 Aircraftman 2nd Class **Frederick Henry Pickering** RAFVR aged 35 was shot when an automatic pistol fired spontaneously during a demonstration at RAF Leighton Buzzard in Stanbridge Road; he died in hospital in Luton. He was the husband of Ivy Pickering of 23 Elsdon Street, Felling-on-Tyne, County Durham and son of Charles and Agnes Pickering of Gateshead, County Durham. Leighton Buzzard Cemetery, Vandyke Road Section KK Grave 16.

1940 THURSDAY 5 DECEMBER: 5954692 Private **Kenneth Ernest Douglas Smith** aged 25 was a traffic control dispatch rider in the Beds & Herts Regiment, accidentally killed on military duty. His motorcycle collided with a roadside pole at Wyton, Huntingsdonshire. Son of Henry and Florence Smith of 37 Vandyke Road, Leighton Buzzard, he had been employed as a grocer's assistant at the International Stores, High Street, Leighton Buzzard. Leighton Buzzard Cemetery, Vandyke Road Section O Grave 325.

1941 TUESDAY 21 JANUARY: Mavis Audrey Williams aged 18 died after 'a solitary German aircraft was heard circling over the village of Heath and Reach'. It dropped a high explosive and some incendiary bombs and most of the latter fell on farmland, doing no damage. Mavis, however, received a direct hit on her shoulder and was killed instantly. The Williams family had been in the Leighton Buzzard area since 1930. A daughter of Mr and Mrs A. Williams of Thomas Street, Heath, she had kept house for her father since leaving school. St Leonard's Churchyard, Heath and Reach Section A Plot 127.

1941 MONDAY 31 MARCH: 60429 Flying Officer **William George Airey** RAFVR aged 35 died at The Bath Club, London W1. He had been living at 'Wragmore', Stewkley Road, Soulbury and was originally from Liverpool. Allerton Cemetery, Liverpool Section 1A C of E Grave 83.

1941 TUESDAY 8 APRIL: 649052 Sergeant **Edward John Cook** RAF aged 19 of 55 Squadron is likely to have been killed in Bristol Blenheim IV T2381, having taken off from El Gubbi in Libya to reconnoitre Benghazi. These are the details available for Flight Sergeant Ernest Patrick Conroy Vignaux, one of the two other airmen who died on the same day; they are buried in adjacent graves. His squadron had been patrolling the Gulf of Suez since September 1939. He was the son of Ernest and Elizabeth Cook of Eaton Bray, Bedfordshire. Knightsbridge War Cemetery, Acroma, Libya Plot 11 Row D Grave 22.

1941 TUESDAY 8 APRIL: 1629751 Gunner **Leslie Walter Turner** aged 20 of 356 Battery 111th Heavy AA Regiment RA died in Scotland. War diaries show that his regiment was defending the major shipyards and docks of Glasgow and the Clyde. He was the son of Walter and Lily Turner of Manor Cottage, Eggington. St Michael's Churchyard, Eggington, north of the church.

1941 WEDNESDAY 17 APRIL: 79077 Flight-Lieutenant **Seymour Fitzclare Clifford** RAFVR aged 38, according to an entry in the probate register, died at Cranmer Court, Chelsea on 17 April 1941. Since joining the RAF, Seymour, and almost certainly his wife Ludmilla, had been billeted at 'Burgate', 17 Stoke Road, Linslade. This is part of the entry for that night from 'the real diaries of Eileen Kelly, a 19-year-old woman living and working in London at the height of the Blitz': 'Wednesday 16th April 1941: Worst Blitz of the War. Land Mine at Cranmer Court.' The 'land mine' would have been a parachute mine. He was the husband of Ludmilla Clifford and son of George and Clare Clifford (of Rangoon and of Frensham, Surrey). St Mary's Churchyard, Old Linslade Section F Grave 76.

1941 SATURDAY 26 APRIL: T/188356 **Henry Percy Sims** aged 25 of 308 Reserve Motor Transport Company, RASC, was in Greece at the time of the Allied evacuation. Unternehmen (Operation) Marita, the long-anticipated German attack against both Greece and Yugoslavia had begun on 6 April. The campaign was hindered by poor communications between the Greek and Allied commanders, by the difficult terrain and by the under-developed road and rail systems. Greek and Allied troops began to fall back as the Germans moved rapidly through the country. By 20 April the Greek government agreed that the force should be evacuated and on 23 April the Greek army itself surrendered. King and government escaped to Crete and the evacuation of Allied forces, begun on 24 April, continued for a week. More than 50,000 British, Australian, New Zealand and Polish troops were evacuated, forced to abandon all of their heavy equipment. Most were taken to Crete, others went to Egypt, but a number of small, isolated groups and individuals were cut off from the retreat and left behind in Greece. He was the husband of Kathleen Sims of 35 Albert Street, Bletchley and son of Arthur and Alice Sims of *The Ewe and Lamb*, 17 Bridge Street, Leighton Buzzard. Athens Memorial, Greece Face 8.

1941 MONDAY 28 APRIL: 2014580 Sapper **Robert Sharratt** aged 21 of 583 Field Company the Royal Engineers (RE) was one of the twenty-one RE men killed when a German bomb hit their air raid shelter at HMS *Raleigh*, a training establishment in Torpoint, Cornwall. Son of Adam and Elsie Sharratt of King's Farm, Edlesborough, he had worked as a carpenter in a family firm, Messrs J. Sharratt & Sons of Eaton Bray, and had played for Eaton Bray Football Club. Torpoint (Horson) Cemetery, Cornwall Grave 64.

1941 FRIDAY 2 MAY: 2069800 Sapper **Thomas Fairfax** aged 20 of 504 Field Company RE died on active service near Haifa Bay in Palestine. Haifa was of great strategic importance during the Second World War because of its deep-water harbour and airfield. It was also the terminus of the railway line from Egypt and of the Kirkuk–Haifa oil pipeline, becoming one of the main supply bases and arms depots serving the Middle East forces. He was the son of Jesse and Gertrude Fairfax of 'Bembridge', Grovebury Road, Leighton Buzzard. Haifa War Cemetery, Israel, Plot D Grave 41.

1941 SUNDAY 11 MAY: James Powell Peasland aged 56 died aboard the barge *Rose* at Paddington Stoplock on the Paddington Branch of the Grand Union Canal. The boat on which he and his son had travelled to London received a direct hit during an air raid. Before retiring because of serious leg injuries, he had worked a horse-drawn canal boat for more than thirty years for Messrs George Garside, Sand Merchants of Leighton Buzzard. He was the husband of Mary Ann Peasland of 34 Old Road, Linslade. St Mary's Churchyard, Old Linslade Plot B119.

1941 WEDNESDAY 21 MAY: 5678423 Private **Bertie Ephgrave** aged 25 of 5th Battalion, The Somerset Light Infantry (Prince Albert's), had been posted to Buckinghamshire and given responsibility for defending airfields. His was an accidental and instantaneous death, caused by the fuselage of a parked aircraft falling on him while he was waiting to go on duty. The plane had made a forced landing some days earlier and had been pushed and fastened into position. His death was registered in the Aylesbury registration district. He was the husband of Annie Ephgrave of 113 Station Road, Church End, Finchley and son of Sidney and Mabel Ephgrave of 21 Ingleway, North Finchley. St Mary's Churchyard, Old Linslade, Section F Grave 77.

1941 SATURDAY 24 MAY: P/KX 97430 Stoker (1st Class) **Percy James Davis** RN aged 26 was on HMS *Hood*. The British Isles were too small to produce all of their own food and Germany knew that controlling the Atlantic would prevent the importing of vital supplies. From 6 to 22 May 1941, *Hood* was based at Scapa Flow, her remit being to cover the North Sea and Atlantic from the threat of German surface raiders. There had been reports that the German battleship *Bismarck* was on the move and from 16 May *Hood* was on alert. At 12.50 a.m. on Thursday 22 May, she and the battleship *Prince of Wales* and six destroyers headed for waters off southern Iceland, ordered to intercept *Bismarck* and the accompanying heavy cruiser *Prinz Eugen*. The following day the two German vessels were sighted between Greenland and Iceland and on 24 May the Battle of the Denmark Strait began. At 5.52 a.m., *Hood* opened fire; *Prinz Eugen* replied. Fire broke out on *Hood* but apparently did not penetrate the engine rooms, since 'her speed remained unaltered to the end'. Only a few minutes later *Hood*'s bow reared up and within three minutes she had sunk, following the explosion resulting from a deep-penetrating hit from *Bismarck* into her after magazines. Out of a crew of 1,418 on board HMS *Hood*, all but three were killed in the explosion or died in the water shortly after it. Son of William and Amy Davis of Watford and a grandson of Charlotte Moxon of 36 Baker Street, Leighton Buzzard, with whom he had lived since the age of 3, he had worked for the Bedford Tile Company in Leighton Buzzard. Portsmouth Naval Memorial, Hampshire Panel 55 Column 1.

1941 SUNDAY 1 JUNE: EX/5309 Royal Marine **Vivian Jack Prudent** aged 19 of Mobile Naval Base Defence Organisation I (MNBDO) disappeared during the evacuation of the last Allied troops from Crete. The function of the MNBDO was to provide the Fleet with a base in any part of the world, whether on the coast of a mainland or an island, within a week, and to defend it when prepared, but there had been time only for advance groups to arrive on Crete. By the evening of 30 May most of the island was in German hands. He was the son of John 'Jack' and Annie Prudent of Leighton Buzzard and a grandson of John and Alice Burchell of 48 Regent Street, Leighton Buzzard, with whom he had lived since early childhood; he had worked in Dunstable and had been a server at St Andrew's church, Leighton Buzzard. Plymouth Naval Memorial, Devon Panel 103 Column 2.

1941 WEDNESDAY 10 SEPTEMBER: 1384380 Leading Aircraftman **John Victor William Shand-Kydd** RAF aged 18 was killed on active service in the Aylesbury area. He was a pilot under training at RAF Hullavington in Wiltshire at the time and was killed in a flying accident. A plane belonging to this school, Master I – T8639 – 9 FTS, dived into the ground in Vasey Field, Horton, near Ivinghoe, Buckinghamshire. The son of Lieutenant-Colonel Norman and Frances Shand Kydd of Horton Hall, Leighton Buzzard, he had joined the RAF on leaving school. St Giles' Churchyard, Cheddington.

1941 FRIDAY 12 SEPTEMBER: 1164381 Sergeant Observer **Richard Louis 'Dick' Fitsell** RAFVR aged 21 of 10 Squadron was the bomb-aimer when, at 8.47 p.m. on Thursday 11 September 1941, Armstrong-Whitworth Whitely V, P5109 took off from RAF Leeming, North Yorkshire. Their target was Warnemünde on the Baltic coast of Germany, the location of the Heinkel aircraft factory, but the aircraft was last heard of on W/T indicating that the crew were preparing to ditch some 80 miles off the east coast of England. No further contact was made and the entire crew were presumed killed, the loss being recorded as occurring on the night of 11/12 September 1941. The son of Henry and Lucy Fitsell of Stanbridge Road, Leighton Buzzard, he had been employed as clerk to the Rating Officer of Wing Rural District Council. Runnymede Memorial, Surrey Panel 43.

Richard Louis Fitsell: 'Dick' Fitsell's family had long military connections; he had joined the ATC while at The Cedars School.

1941 SATURDAY 22 NOVEMBER: 5953391 Private **Francis John King** aged 29 of 1st Battalion Beds & Herts Regiment died in hospital in Tobruk, Libya, during Operation Crusader, relieving the Siege of Tobruk 18 November–30 December 1941. Earlier in 1941 the battalion had sailed to the Middle East and by November were heavily involved in the Battle for Tobruk, fighting against Rommel's Afrika Corps. Tobruk was the only deep-water port in Eastern Libya and its capture was essential for an advance on Alexandria and Suez. He was the son of George and Lottie King of 14 St Andrews Street, Leighton Buzzard. Tobruk War Cemetery, Libya Plot 8 Row A Grave 10.

1941 SUNDAY 23 NOVEMBER: 5948112 Sergeant **William Charles Ward** aged 27 of 1st Battalion Beds & Herts died during Operation Crusader. Husband of Letitia Ward of Salford, Lancashire and son of the late Francis Ward and of (Susan) Louisa Ward, later Eastaff, of Leighton Buzzard, he had been with the army in Egypt since 1936. Knightsbridge War Cemetery, Acroma, Libya Plot 11 Row H Grave 10.

1941 TUESDAY 23 DECEMBER: 1164131 Aircraftman (1st Class) **William James King** RAFVR of 67 Squadron died aged 21 in Burma. On 23 December, Japanese bombers sortied against Rangoon. The Imperial Japanese Army Air Force formation was intercepted by RAF Brewster Buffalos of 67 Squadron, which shot down eight Japanese planes. Rangoon and Mingaladon airfields were successfully bombed, two Buffalos and two P-40s being destroyed on the ground, and one P-40 crashing when attempting to land on a bomb-damaged runway. He was the son of George and Nellie King of Leighton Buzzard. Rangoon War Cemetery, Myanmar, Plot 3 Row B Grave 13.

1942 SATURDAY 14 FEBRUARY: 35984 Major **William Haslam Merry** aged 39 of the Bedfordshire Yeomanry (148th Field Regiment RA), having been captured in Singapore with his regiment, was killed while on his way to establish what had happened to the men under his command. He was the husband of Mary Merry of 'High Meadows', Aylesbury Road, Wing, Buckinghamshire and son of Arthur and Alice Merry of Wing. Prior to the outbreak of war, he had worked in the family business of Stafford, Rogers and A.W. Merry, auctioneers and estate agents, in Leighton Buzzard and had for several years commanded the Dunstable and Leighton Battery of the Bedfordshire Yeomanry. Kranji War Cemetery Plot 34 Row A Grave 20.

In early 1939 a recruiting drive for the Territorial Army took place in Parson's Close recreation ground, Leighton Buzzard. Major William Haslam Merry encouraged the local men to enlist.

1942 SATURDAY 14 FEBRUARY: 895402 Gunner **Norman Royston Norman** aged 22 of 419 Battery the Bedfordshire Yeomanry died during the Japanese capture of Singapore. He was the husband of Joyce Norman of 'The Black Horse', 21 North Street, Leighton Buzzard and the son of Charles and Elsie Norman. Singapore Memorial Column 38.

1942 SUNDAY 15 MARCH: P/K Petty Officer Stoker **Thomas Alves** RN of HMS *Vortigern* died aged 47 when his ship was lost with all hands off the Norfolk coast. *Vortigern* was escorting coastal convoy FS-749 when she was sunk off Cromer at 55°06'N, 01°22'E by the German motor torpedo boat (E-boat) S-104. The wreck, designated a protected place, lies in 17m of water at 53°13'06'N, 01°06'54'E. He was the husband of Alice Alves of Hornsey and son of John and Mary Alves of Midlothian, Scotland. Thomas had been a coal-miner before enlisting in the army in November 1914 in Great Yarmouth and served in the Royal Garrison Artillery during the First World War, marrying in Hampshire in 1919. His name is on the war memorial in the village, but his connection with the area around Mentmore is not known. Portsmouth Naval Memorial Panel 67 Column 3.

1942 SUNDAY 5 APRIL: Surgeon Lieutenant **Stanley Walter Pratt** (MB BS MRCS LRCP) Royal Navy Volunteer Reserve (RNVR) aged 29 was one of the ship's complement of HMS *Dorsetshire*. Since the previous December, *Dorsetshire* had been assigned to the Eastern Fleet for convoy work in the Indian Ocean. At the beginning of April 1942, *Dorsetshire* and her sister ship *Cornwall* were detached from the fleet to escort the aircraft carrier HMS *Hermes* to Trincomalee at Ceylon for repairs, 'hunting for surface raiders and an enemy fleet'. On 4 April the Japanese carrier fleet was spotted. The two cruisers left the harbour and, after a hurried refuelling at sea, set out for Addu Atoll shortly after midnight. The following day they were sighted by a spotter plane from the Japanese cruiser *Tone* about 200 miles south-west of Ceylon. During this, the Easter Sunday Raid, a wave of Aichi D3A Val dive-bombers from Japanese carriers attacked and sank the two ships. It is not known whether Lieutenant Pratt died on the ship or in the sea after it had sunk. However, an eyewitness from the crew reported that 'the sick bay … was caught by a direct hit … killed everyone in the sick bay bar two of the doctors.' Another wrote that:

> One of the repair parties on an inspection round had no sooner gone through the doorway at the end of the sick bay than there were two terrific explosions from somewhere aft … we got two direct hits … These waves of bombers were followed by fighters, which dived, machine-gunning the decks and gun stations and shattering most of the lifeboats with explosive bullets … It all happened in

a few minutes … A bomb came clean through the deck and passed through the dispensary at the end of the sick bay and exploded in the marines' mess deck.

Son of Walter and Annie Rosa Pratt of 'Cassio', 36 Hockliffe Street, Leighton Buzzard, he had qualified in medicine at University College, London and had held hospital appointments in London and Leicester. Plymouth Naval Memorial, Devon Panel 76 Column 3.

1942 SUNDAY 5 APRIL: CH/X 1970 Marine **Ernest George Sifleet** aged 22 of the Royal Marines (RM) was among the casualties reported as 'missing in the Indian Ocean' a few miles away from HMS *Dorsetshire* when his ship HMS *Cornwall* suffered a similar fate. *Cornwall* was hit eight times and sank bow first a few minutes later. He was the son of Sydney and Mary Sifleet of 56 North Street, later 33 Stanbridge Road, Leighton Buzzard. Chatham Naval Memorial, Kent Panel 65 Column 2.

1942 THURSDAY 7 MAY: 1216130 Sergeant Pilot **Harold Ernest Perry** RAFVR aged 21 based at RAF Edgehill died when Wellington T2824 of 21 OTU crashed on the Stratford road just above the village of Balscote while undertaking night circuits and landings. He was the son of Albert and Gertie Perry of 4 Bedford Street, Bletchley. Bletchley Cemetery New Part Grave 1042.

1942 WEDNESDAY 10 JUNE: 5954560 Private **Aubrey Frederick Robert Hardy** aged 26 of 4th Battalion The Buffs (Royal East Kent Regiment) was in Malta. During the convoy battles there were overhead air skirmishes, rescue launches rushing to sea and gunfire from the ground. The island was in a constant state of battle. He was the son of Frederick James Hardy and Florence Hardy of Bower Lane, Eaton Bray. Imtarfa Military Cemetery, Malta, Plot 3 Row 3 Collective grave 2.

1942 SATURDAY 27 JUNE: T/156952 Driver **George Oliver Draper** RASC was killed instantly aged 24 while driving his Norton motorcycle along a main road near Woking, Surrey: the machine struck a concrete road block at a double bend in the road. Eight men of the King's Royal Rifles in the charge of a sergeant formed the bearer party at his funeral. He was the husband of Mary Draper of Thrift Road, Heath and Reach, son of Ernest and Sarah Scotchings of 7 St Andrews Street, Leighton Buzzard. Leighton Buzzard Cemetery, Vandyke Road Section K Grave 171.

1942 THURSDAY 3 SEPTEMBER: 131880 Pilot Officer **Charles Frederick Walter Underwood** RAFVR aged 24 of 61 Squadron was a member of the crew of Avro Lancaster W4136, which failed to return after setting out at 11.10 p.m. on 2 September from RAF Syerston, Lincolnshire. It was shot down in the night, part of a strong force of RAF bombers raiding Karlsruhe, one of the chief cities of the Rhineland. This was an important transport centre and junction of the east–west railway from Munich and south-west Germany to Strasbourg and France, and also of the north–south line from Mannheim and the Ruhr to Italy and Switzerland. He was the son of Frederick and Edith Underwood of 23 Durbar Road Luton, originally of Leighton Buzzard and Linslade. Runnymede Memorial, Surrey Panel 72.

1942 FRIDAY 18 SEPTEMBER: Frank Thomas Clarke aged 12 died instantly when a grenade found at a barricade in woods near his home exploded. Some lads had found some empty cartridge cases and an unexploded No. 69 hand grenade, left during military exercises, and had brought these back to the village. He was the son of Frank and Alice Clarke of Heath. St Leonard's Churchyard, Heath and Reach.

1942 FRIDAY 2 OCTOBER: D/K 65312 Petty Officer Stoker **Charles John Morse** RN aged 36 had been stationed at shore station HMS *Tamar*, the RN's base in Hong Kong, which had been an actual ship scuttled by the British in Kowloon Harbour on 12 December 1941, after which a shore establishment, a 'stone fortress' was used. On Christmas Day 1941, the Japanese had landed and taken prisoners, among them Charles Morse. On 27 September 1942 the *Lisbon Maru* left Hong Kong for Shanghai with 1,816 prisoners onboard. She was torpedoed on 1 October by USS *Grouper*, 6 miles (9km) from Tung Fusham Island, off the Chinese coast, 29°57'N, 122°56'N. The Japanese soldiers were seen to abandon ship but would have battened down the hatches before embarkation, imprisoning the British POWs. The roll shows the names of 820 men, including Charles Morse, who were not seen alive after 2 October. He was the husband of Winifred Morse of Leighton Buzzard and son of William and Mary Morse of Gloucestershire. Plymouth Naval Memorial, Devon Panel 69.

1942 FRIDAY 30 OCTOBER: 5505188 Private **Percy Clifford Marks** aged 26 of 5th Battalion Royal Sussex Regiment (Cinque Ports) was killed in action in the Middle East. He was a motor transport driver and had 'volunteered for what he knew to be dangerous work'. His battalion was penned down in a very difficult position by a strong enemy force and the only way food could be transported to some of the men was by carrier. The carrier he was driving was hit by shellfire, killing him instantly. The husband of Ethel Marks of Sandown, Isle of Wight, son of Florence Marks and stepson of William Alfred Quick of Heath Road, he had been an apprentice linotype operator in the *Leighton Buzzard Observer* office before leaving the area to work for an Isle of Wight newspaper. El Alamein War Cemetery, Egypt, Plot 17 Row D Grave 25.

1942 MONDAY 2 NOVEMBER: 108557 Flying Officer **Thomas William Lound Miller** RAF of 2842 Squadron was stationed at RAF Leighton Buzzard and died aged 48 of a heart attack at RAF Dagnall. Also known as RAF Edlesborough, this was a recently created wireless transmitting station on Gallows Hill, Edlesborough, working alongside RAF Leighton Buzzard and RAF Stoke Hammond in Dorcas Lane (the receiving station). Though neither a local nor commemorated on any local war memorial, he was buried in Leighton Buzzard. The informant of his death was his second wife. Whether she was living in this area permanently or temporarily is unknown – they had married in Scotland and lived in Wales. He was the husband of Helen Webster Miller of Abergele, Denbighshire and son of Donald and Elizabeth Miller of Norfolk. Leighton Buzzard Cemetery, Vandyke Road Section P Grave 16.

1942 TUESDAY 10 NOVEMBER: 5957832 Private **George Edward Hewitt** of 1/7th Battalion, the Queen's Royal Regiment (West Surrey) died aged 31 in hospital of wounds received in action and was buried in Alexandria. His battalion was with the 44th (Home Counties) Division at El Alamein. The son of William and Caroline Hewitt of 1 Baker Street, Leighton Buzzard, he had worked for Mr Chandler, coal merchant, in Linslade. Alexandria (Hadra) War Memorial Cemetery, Egypt Plot 4 Row B Grave 18.

1942 TUESDAY 24 NOVEMBER: T/187195 Driver **Owen Percy Maunders** RASC aged 29, a dispatch rider stationed in Scotland, died in hospital after five days' illness from a ruptured appendix and peritonitis. The husband of Dorothy Maunders of Ayr and the son of Walter and Cissie Maunders of Edlesborough, he had been a letterpress printer at Waterlow & Sons Ltd, Dunstable and a member of Eaton Bray Tennis and Cricket Clubs. Ayr Cemetery, Ayrshire Section S 1931 Division Grave 3110.

1942 THURSDAY 24 DECEMBER: Lieutenant **Richard Arthur Edmunds** RNVR aged 31 of HM Submarine *P 216* (later HMS *Seadog*) was reported as missing, presumed killed, in the Barents Sea, off North Cape (Nordkapp) on the northernmost tip of Norway. He was the husband of Phyllis Edmunds and the son of Charles and Eleanor Edmunds of Ledburn Manor House. Portsmouth Naval Memorial Panel 71 Column 1.

1943 TUESDAY 12 JANUARY: 5952560 Private **John 'Jack' Bond** aged 23 of 5th Battalion Beds & Herts died of dysentery at Tarsao No. 1 Camp while a

POW, working on the Thailand–Burma railway. He was first buried at Tarsao but was later reinterred. He was the husband of Winifred Bond of 7 Reach Green, Heath and Reach. Kanchanaburi War Cemetery Plot 4 Row F Grave 10.

1943 SUNDAY 17 JANUARY: 307696 Sergeant (Wireless Operator) **Alan Richard Saunders** RAFVR of 76 Squadron was killed aged 27 when Halifax II – DT 647 coded MP-P Bomb Op, which had taken off at 4.22 p.m. from Linton-on-Ouse, target Berlin, was lost without trace off the coast of Holland, 70 miles (45km) north-west of Juist. They had been shot down at 9.55 p.m. by German aircraft NJG3; Oberleutnant Paul Zorner, who was to be one of the most famous flying aces of the Luftwaffe, had claimed his first victory. Saunders was the son of Frederick and Florence Saunders of 7 Manor Road, Cheddington. Runnymede Memorial Panel 163.

1943 MONDAY 25 JANUARY: 1922320 Sapper **Marshall Claude Hedges** RE aged 23, attached to 30th Battalion the East Yorkshire Regiment (The Duke of York's Own) died after a short illness in Driffield, Yorkshire. He was the son of Arthur and Emily Hedges, of 73 Tilsworth Road, Stanbridge. St. John the Baptist Chapelyard, Stanbridge north-east corner.

1943 TUESDAY 16 FEBRUARY: 5831314 Private **John William Whiting** aged 27 of the 2nd Battalion the Cambridgeshire Regiment died in Thailand, while a POW working on the Thailand–Burma Railway. He was the son of Frederick and Rosa Whiting of Leighton Buzzard. Chungkai War Cemetery Plot 12 Row F Grave 12.

1943 FRIDAY 5 MARCH: 2146752 Driver **William Charles Dyer** aged 29 of 108th Field Park Company, RE died in the General Hospital, King's Lynn, Norfolk a week after a serious accident while on military duty collecting scrap metal – an object, thought by those concerned to be a lorry silencer, exploded unexpectedly and he sustained severe injuries. He was the husband of Ada Dyer of 5 The Square, Southcourt, Linslade and the son of Henry and Sarah 'May' Dyer of Linslade, formerly of Bermondsey. St Mary's Churchyard, Old Linslade Section G Grave 73.

1943 WEDNESDAY 17 MARCH: 149113 Lieutenant **Arthur Geoffrey Buchanan** aged 23 of No. 7 Anti-Tank Platoon, 6th Battalion Grenadier Guards was wounded by a mine and killed by mortar fire during a nocturnal attack on a horseshoe feature on the Mareth Line. The Mareth Line was a system of French-built fortifications in southern Tunisia which had fallen into Axis hands. He had been living in Ascott Cottage, Wing and was the son of Captain John

Nevile Buchanan DSO, MC, and Nancy Buchanan of Bledlow Ridge. Sfax War Cemetery, Tunisia Plot 2 Row BB Grave 4.

1943 SATURDAY 3 APRIL: 1216729 Sergeant (Air Gunner) **George Edward Gadd** RAFVR aged 20 of 101 Squadron was killed by flak on-board Avro Lancaster W4923 while crossing the Dutch coast after bombing Essen. The pilot managed to fly back to base at Holme on Spalding Moor in Yorkshire and George's death was registered there. The son of George and Emily Gadd of 18 High Street South, Dunstable, he had worked for a Luton typewriter firm and then for A.C. Sphinx Sparking Plug Company at Watling Street, Dunstable. Dunstable Cemetery Section K Grave 546.

1943 MONDAY 5 APRIL: 5957929 Private **Kenneth George Rollings** aged 23 of the Beds & Herts Regiment died of tuberculosis at Papworth Village Settlement near Papworth Everard in Cambridgeshire. He was the husband of Iris Rollings of Dunstable, son of Eva and stepson of Charles Cook. Dunstable Cemetery Section K Grave 547.

1943 WEDNESDAY 7 APRIL: 5344466 Private **Henry Richard Baines** aged 27 of 1st Battalion the Royal Berkshire Regiment was reported missing in India – more specifically, his battalion had been involved in the first tentative Allied attack into Burma, fighting in the western coastal region. It is probable that he died during an attempt to ambush the Japanese on the Maungdaw–Buthidaung road, which proved beyond the capability of the tired troops. Husband of Minnie Baines of 12 Crown Terrace, Kentish Town and son of William and Rose Baines of 16 East Street, Leighton Buzzard, he had worked for the Marley Tile Company in Stanbridge Road, Leighton Buzzard. Rangoon Memorial, Myanmar Face 15.

1943 SATURDAY 17 APRIL: 5951937 Private **Walter George Smy** aged 32 of 5th Battalion Beds & Herts died as a POW in Wampo North Camp on the banks of the River Kwai. He was the husband of Ethel Smy of 41 Ashwell Street, Leighton Buzzard and son of John and Margaret Smy of Orford, Suffolk. Singapore Memorial Column 64.

1943 FRIDAY 23 APRIL: 947194 Driver **George Thomas Gibbins** aged 21 of The Warwickshire Yeomanry, the Royal Armoured Corps (RAC), died from his injuries in North Africa. The son of Charles and Jessica Gibbins of 110, Soulbury Road, Linslade, he had worked for Bagshawe & Co Ltd of Church Street Dunstable. Beirut War Cemetery Plot 3 Row C Grave 3.

1943 SATURDAY 24 APRIL: 321004 Lance Corporal **Percy Harold Hunt** aged 23 of the Queen's Bays (2nd Dragoon Guards), RAC, was killed in action in Tunisia. The tanks of the Queen's Bays, their sand-colour changed to dark green, had been loaded onto transporters near Sfax and by 23 April were concentrated 3 miles south of Goubellat. Moving out with the 9th Lancers, they had advanced 5.5 miles (9km) by dusk, delayed principally by mines that had been laid in standing corn and were thus very difficult to detect. Six Crusaders and two Shermans were damaged. He was the husband of Doris Hunt of 31 Beale Street, Dunstable and the son of Henry and Ada Hunt of Northall. Medjez-El-Bab War Cemetery, Tunisia Grave 2 E 7.

1943 TUESDAY 11 MAY: 1605194 Leading Aircraftman **Reginald R.I. Smythe** aged 29 RAFVR Air Gunner under training at Air Gunnery School, RAF Pembrey, died in Wales, his death being registered in Llanelly registration district. On that date Blenheim IV Z6348 of 1 AGS (Air Gunnery School) dived into the ground at Pant Teg Farm, Pinged, a mile east of Pembrey in Carmarthenshire after the port flap failed. The husband of Daphne Smythe of Ivinghoe and the son of Frank and Sarah Smythe. St Mary the Virgin, Ivinghoe Row 9 Grave 1.

1943 SATURDAY 29 MAY: 619819 Corporal **John Edward Steadman Deverell** RAF aged 24 of No. 255 Group was based in India. One other ground crew member died on the same day and is buried in the same cemetery. Deverell was the son of Edward and Dorothy Deverell of Home Farm Lodge, Park Road, Tring, Hertfordshire. Kirkee War Cemetery, India Plot 2 Row F Grave 11.

1943 SATURDAY 5 JUNE: 2369664 Signalman **Samuel Henry Dawson** aged 38 of the Royal Corps of Signals died in Yorkshire, his death recorded in Bulmer registration district. He was the husband of Dorothy Dawson of Pitstone and the son of Hugh and Emma Dawson of Staffordshire. St Giles Churchyard, Cheddington.

1943 THURSDAY 24 JUNE: 5951976 Private **Reginald Thomas Cox** aged 23 of 5th Battalion Beds & Herts died at Takanon of beriberi and dysentery, while working as a POW on the Thailand–Burma railway. He was buried first at Takanon, the camp where he had died, and was later re-interred. He was the son of Thomas and Annie Cox of 3 Council Houses, Leighton Road, Eggington. Kanchanaburi War Cemetery, Thailand Plot 2 Row M Grave 78.

1943 WEDNESDAY 30 JUNE: 5832480 Private **Albert Victor Blew** aged 30 of 5th Battalion Suffolk Regiment was with his regiment in Singapore when it was seized by the Japanese. Like so many other POWs he was sent to work building the Burma–Thailand railway. He was the husband of Winifred Blew of 76 Ashwell Street, Leighton Buzzard and son of Margaret Blew of Wing. Kanchanaburi War Cemetery, Thailand Plot 10 Row F Collective Grave 2-10 L 4.

1943 THURSDAY 8 JULY: 1172593 Flight Sergeant Navigator **Eric Vernon Frederick Weatherley** RAFVR aged 21 died in Libya. A relative of the other RAF man who died the same day and was buried in the same place discovered that 'He had been in a Marauder which crashed on landing when the undercarriage failed.' He was the son of Charles and Lilian Weatherley, temporarily of Croxley Green, Hertfordshire but from Leighton Buzzard. Tripoli War Cemetery, Tobruk Al Butnan, Libya Plot 11 Row A Grave 9.

1943 SUNDAY 18 JULY: 1633497 Gunner **Herbert Thomas Clayton** aged 33 of 241 Battery 77th Heavy AA Regiment RA died as a Japanese POW, in all probability while working on the Thailand–Burma railway. His regiment had been captured in Java in March 1942 and transported to work elsewhere. He was the husband of Hilda Clayton of Summer Street, Leighton Buzzard and the son of Jesse and Alice Clayton. Kanchanaburi War Cemetery, Thailand Plot 8 Row E Grave 35.

1943 SUNDAY 1 AUGUST: 900696 Bombardier **Kenneth Harold Gilbert** aged 24 of 419 Battery the Bedfordshire Yeomanry had been captured with his regiment in Singapore on 14 February 1942; he died of beriberi and dysentery at Kinsayok as a Japanese POW, working on the Thailand–Burma railway. He was the son of Harold and Maude Gilbert of 'Chalmers', Billington Crossing, Leighton Buzzard. Kanchanaburi War Cemetery, Thailand Plot 8 Row G Grave 37.

1943 TUESDAY 10 AUGUST: 918243 Gunner **James Arthur Edwin Proctor** aged 22 of 512 Battery the Bedfordshire Yeomanry was captured with his regiment in Singapore on 15 February 1942. He died of beriberi in Tarsao. The husband of Doris Proctor of 65 Granville Road, Southtown, Great Yarmouth, Norfolk and son of James and Constance Proctor of 130 Soulbury Road, Linslade, he had been apprenticed to Horace Brantom, builder, of Vandyke Road, Leighton Buzzard. Kanchanaburi War Cemetery, Thailand Plot 4 Row A Grave 74.

1943 SUNDAY 22 AUGUST: 62311 Flight Lieutenant (Pilot Officer) **Arthur Charles Griffin** RAFVR of 201 Squadron was killed aged 27 with seven other members of his crew on-board Short Sunderland DD848 when it crashed on Mount Brandon, on the Dingle Peninsula in Kerry, one of several Sunderlands sent to Biscay that day. Six officers of the RAF attended his memorial service, which was held in the Baptist Chapel, Hockliffe Street, Leighton Buzzard, timed to coincide with his funeral service in Ireland. He was the son of Charles and Emma Griffin of 27 Queen Street, Leighton Buzzard and had been employed by the firm of I. & R. Morley Ltd, clothing manufacturers. Irvinestown Church of Ireland Churchyard, Fermanagh, N Ireland Plot 1 Grave 12.

1943 WEDNESDAY 8 SEPTEMBER: 6013705 Private **James Robert Mordecai** aged 26 of 2/5th Battalion the Essex Regiment died in Italy. He was the husband of Irene Mordecai of Eaton Bray and the son of Herbert and Elizabeth Mordecai, latterly of Eaton Bray. Cassino Memorial Panel 8.

1943 WEDNESDAY 8 SEPTEMBER: 136933 Flying Officer **Leslie Frederick Smith** DFC RAFVR died in hospital in Newmarket of his injuries after being shot down by a night intruder, ME110, over Great Thurlow on the night of 7 September 1943. Based with 1657 Heavy Conversion Unit (HCU) at Stradishall, he was giving night-flying instruction to a Sergeant Gilkes and his crew in Stirling W7455. Husband of Norah Smith of 'Killala', Stoke Road, Water Eaton and son of James and Florence Smith of Wharf House, Water Eaton, he had been employed in the family milling business. Bletchley Cemetery, Buckinghamshire New Part Grave 1061.

1943 FRIDAY 10 SEPTEMBER: 10695541 Driver **Kenneth Roland Rayner** aged 38 of 56th Division Signals, Royal Corps of Signals. The Allied invasion of Italy had begun on 3 September 1943 and he was driving his officer and a sergeant to a battalion that had been cut off when they were attacked by a tank. The husband of Helen Rayner of 4 Wing Road, Linslade and son of Peter and Eliza Rayner of Adstock, Buckinghamshire, he had been employed by William Allder, hay and chaff merchant. Salerno War Cemetery Plot 2 Row B Grave 36.

1943 WEDNESDAY 20 OCTOBER: 1387726 Flight Sergeant (Navigator) **Derrick Leslie Walter Horn** RAFVR aged 20 of 115 Squadron, based at Little Snoring, Norfolk, was on Avro Lancaster DS769, which was lost on a raid to Leipzig during the night of 18/19 October 1943, crashing at Engersen. The German invasion of France had triggered the bombing campaign against Germany, and 115 Squadron remained part of the main bomber force until the end of the war. The son of Walter and Elsie Horn of High Street, Ivinghoe, he had joined the RAF upon leaving school. Berlin 1939–1945 War Cemetery, Germany Plot 9 Row E Joint Grave 7-8.

1943 THURSDAY 28 OCTOBER: 1793795 Gunner **Reginald Victor Day** of 253 Battery, 81st Heavy AA Regiment RA was killed aged 22 on a troop ship that was dive-bombed in the Suez Canal. He was the husband of Alice Day of Bournemouth, Hampshire and the son of James and Ethel Day of Eaton Bray. Brookwood Memorial Panel 3 Column 1.

1943 TUESDAY 2 NOVEMBER: 2075596 Sapper **Arthur William Murgett** aged 22 of 251st Field Park Company, RE died as a POW, probably while working on the Thailand–Burma railway. His regiment, having arrived in Singapore without transport or equipment on 5 February 1942 after being shipwrecked when the *SS Empress of Asia* was bombed by the Japanese, was taken prisoner on 14 February. The son of Herbert and Helena Murgett of Billington Manor Cottages, Great Billington, he had been an apprentice at Commer Cars, Biscot Road, Luton. Kanchanaburi War Cemetery, Thailand Plot 2 Row B Grave 25.

1943 TUESDAY 23 NOVEMBER: 6019999 Private **George Reginald Washington** aged 26 was one of the twenty-three men of 5th Battalion, the Essex Regiment who died in action somewhere in the central Mediterranean theatre on this date. The Allied force had fought its way up the Adriatic coast and was preparing to attack the Sangro River positions in Italy. The son of Edward and Ada Washington of 55 Plantation Road, Leighton Buzzard, he had been employed by Messrs Boots. Sangro River War Cemetery, Italy Plot 17 Row D Grave 11.

1943 MONDAY 29 NOVEMBER: 1808716 Lance Bombardier **William John 'Jack' Whitman** aged 29 of 239 Battery, 77th Heavy AA Regiment RA was taken as a POW in Java and died in a torpedo attack on the *Suez Maru*. The Japanese had decided to ship the sick back from Ambon Island to Java. Among those on-board the 4,645-ton passenger-cargo ship *Suez Maru* in two of the holds were 422 sick British prisoners. Escorted by a minesweeper W-12, the *Suez Maru* set sail from Port Amboina but while entering the Java Sea and about 327km east of Surabaya, Java, it was torpedoed by an American submarine. Water poured into the holds, drowning hundreds but many managed to escape and swam away from the sinking ship. The Japanese mine sweeper W-12 picked up the Japanese survivors, leaving between 200 and 250 men in the sea. At 2.50 p.m. the minesweeper opened fire, using a machine gun and rifles, after which rafts and lifeboats were rammed and sunk by the W-12. The firing did not cease until all the prisoners had been killed. Husband of Kathleen Whitman of Linslade, formerly of 32 Hartwell Crescent, Leighton Buzzard and son of William and Lottie Whitman of 15 Billington Road, Leighton Buzzard, he had worked for

Messrs Stafford, Rogers and A.W. Merry, auctioneers and had captained the Leighton Buzzard Hockey Club. Singapore Memorial, Singapore Column 11.

1943 THURSDAY 2 DECEMBER: 791518 Lance Bombardier **Kenneth Walter Olney** aged 21 of 131 Battery, 47th Light AA Regiment RA died in Italy. In one of the early raids a bomb burst on or near his billet and he was killed instantly by the blast. 'He died whilst his gun was firing at the enemy, doing his duty without hesitation at a time of great stress.' The son of William and Ethel Olney of Plantation Road, Leighton Buzzard, he had worked for Joseph Arnold & Sons and later for Bernard Sunley & Co., a company specialising in airfield construction. Bari War Cemetery, Italy Plot 14 Row B Grave 34.

1943 WEDNESDAY 15 DECEMBER: 5951354 Private **Joseph Cato** aged 25 of 5th Battalion Northamptonshire Regiment died during the Italian campaign; his regiment formed part of the 11th Infantry Brigade. He was the son of Mark and Annie Cato of the Lock House, Three Locks, Soulbury. Sangro River War Cemetery Plot 9 Row B Grave 13.

1943 THURSDAY 23 DECEMBER: 1577648 Flight Sergeant **Glyn Hankins** RAFVR of 261 Squadron was killed aged 20 whilst serving as a fighter pilot in India. He had been the first member of the Leighton Buzzard ATC Squadron to gain pilot's wings; at the time of his death he was likely to have been flying in a Hawker Hurricane IIC, escorting Douglas Dakotas that were dropping air supplies to forward troops in Burma. Since January 1943 his squadron had been at the Burma front, where it flew both bomber escort and ground attack missions. The son of William and Annie Hankins of 46 Windsor Street, Bletchley, he had joined the RAFVR straight from school. Singapore Memorial, Singapore Column 425.

1943 TUESDAY 28 DECEMBER: 13111706 Private **Henry George Huxley** of the Pioneer Corps died in Banstead Military Hospital, Surrey. Private Huxley's family worked boats on the Grand Union Canal. He was the husband of Louisa Huxley of Linslade and the son of Samuel and Annie Huxley. St Mary's Churchyard, Old Linslade Section F Grave 78.

1944 TUESDAY 1 FEBRUARY: 5955850 Private **James Theodore Sear** aged 23 of 5th Battalion Beds & Herts died in captivity in Tarsao Camp No. 2, his regiment having been captured in Singapore by the Japanese on 14/15 February 1942. He was initially buried in Tarsao No. 2 cemetery, grave number 572. He was the husband of Violet Sear of 3 Star Cottages, Bierton and the son of Leonard and Florence Sear of 34 Hockliffe Road, Leighton Buzzard. Kanchanaburi War Cemetery, Thailand Plot 4 Row E Grave 32.

1944 TUESDAY 1 FEBRUARY: 1306483 Flight Sergeant **Richard White** RAFVR aged 33 was stationed in Leighton Buzzard with 2841 Squadron and died in the area covered by Aylesbury registration district. He was the husband of Phyllis White of 161 Stanbridge Road, Leighton Buzzard son of Arthur and Martha White. Halton St Michael Churchyard Plot 5 Row B Grave 149.

1944 SATURDAY 12 FEBRUARY: D/JX 424270 Ordinary Signalman **Stanley Robert Goodway** RN aged 19 of HMS *Lanka*, an RN shore base in Ceylon, was in fact a member of the crew on the troopship *SS Khedive Ismail*, which had left the African port of Kilindini as part of a convoy bound for Burma. A Japanese submarine fired two torpedoes into the engine-room, blowing up the boilers, and the ship took just two minutes to sink in the Indian Ocean. Further fatalities were caused when British destroyers attacked the submarine by dropping depth charges which killed many survivors who were in the water. He was the son of Albert Cyril and Florence Goodway of Wing. Plymouth Naval Memorial Panel 88 Column 3.

1944 SUNDAY 20 FEBRUARY: 1258560 Corporal **Ernest Hector Gill** RAFVR was killed aged 37 in an explosion in HMS *LST 305* (Landing Ship Tank) off Anzio 41° 14'N, 12° 31'E, which was hit and sunk by a GNAT (German Navy Acoustic Torpedo) fired from U-boat *U-230*. He was the husband of Gwendoline Gill of 12 Victoria Road, Hitchin, formerly of Mill Road, Leighton Buzzard and a nephew of Thomas Ernest and Mary Ellen (Polly) Gill of Prees, Shropshire. He had been under-manager at the Home and Colonial Stores in High Street, Leighton Buzzard. Malta Memorial, Malta Panel 15 Column 2.

1944 TUESDAY 22 FEBRUARY: 256442 Lieutenant **Howard Harvey Noel Heald** aged 21 of the Dorsetshire regiment, attached to The Queen's Royal Regiment (West Surrey) died in Italy during the Anzio campaign. He was the son of Thomas and Ethel Heald of 'Woodville', Woodside, Aspley Guise. Anzio War Cemetery Plot 3 Row X Grave 6.

1944 SATURDAY 11 MARCH: 11406390 Gunner **William John Wesley** aged 31 of 8 Battery, 4th Light AA Regiment died in Kent, his death being registered in Bridge registration district. He was the husband of May Wesley and the son of Frank and Edith Wesley of Cheddington. St Giles Churchyard, Cheddington.

1944 WEDNESDAY 15 MARCH: 5948007 Private **Leslie Charles Ward** aged 23 of 2nd Battalion Beds & Herts died in Italy on the first day of the Third

Battle of Cassino. He was the son of Charles and Florence Ward of 74 Princes Street, Dunstable. Minturno War Cemetery Plot 7 Row B Grave 18.

1944 SUNDAY 9 APRIL: 858909 Lance Sergeant **Frank Haworth Hucklebridge** aged 26 of the Bedfordshire Yeomanry died of amoebic dysentery in Tarsao camp while working as a POW following the capture of his regiment in Singapore in February 1942. The son of Frank and Mabel Hucklebridge of ' The Warren', Hockliffe, he had been apprenticed as a linotype operator on the staff of the *Leighton Buzzard Observer*. Kanchanaburi War Cemetery Plot 4 Row E Grave 28.

1944 TUESDAY 25 APRIL: 139700 Flying Office (Navigator) **Reginald Francis Weedon** RAFVR aged 23 of 100 Squadron took off from Grimsby on his ninth operational flight at 2216 in Avro Lancaster ND328 HW-N bound for Karlsruhe. The plane was reported to have crashed near St Martens-Voeren, 21km north-east of Liège. All the crew were originally buried by the Luftwaffe at St Truiden and after the war their bodies were exhumed and taken to the Heverlee War Cemetery. Husband of Naomi Weedon of Luton and son of Frederick and Florence Weedon of 'Glendale', 130 Waller Avenue, Luton, formerly of Eaton Bray, he had been employed in the statistics office of the Davis Gas Stove Company, Luton. Heverlee War Cemetery, Leuven, Vlaams-Brabant, Belgium Plot 5 Row F Grave 5.

1944 THURSDAY 27 APRIL: 773674 Lance Corporal **Thomas Henry Maynard** aged 33 of 1st Battalion, Northamptonshire Regiment was killed in action on the Indo-Burma frontier. The husband of Doris Maynard of 36 Broomhills Road, Leighton Buzzard and the son of Thomas and Elizabeth Maynard, he had been a regular soldier since 1931, formerly in the Beds & Herts, and had been wounded at Dunkirk. Rangoon Memorial, Myanmar Face 15.

1944 WEDNESDAY 3 MAY: 79156 Captain **Richard Glencairn Boyd-Thomson** aged 27 of 99th (The Royal Buckinghamshire Yeomanry) Field Regiment RA died in Burma. During May-June 1944 his regiment fought in the Battle of Kohima, a decisive victory forcing the Japanese for the first time into retreat across Burma and South-East Asia. The battle took place in three stages. From 3 to 16 April, the Japanese attempted to capture Kohima Ridge, a feature that dominated the road by which the besieged British and Indian troops of IV Corps at Imphal were supplied. By mid-April, the small British force at Kohima was relieved, and from 18 April to 13 May, British and Indian reinforcements counter-attacked to drive the Japanese from the positions they had captured. The Japanese abandoned the ridge at this point, but continued to block the Kohima–Imphal road. He was the son of Glencairn and Doris Boyd-Thomson of Soulbury, Buckinghamshire. Rangoon Memorial, Taukkyan War Cemetery, Myanmar Face 3.

1944 MONDAY 8 MAY: 1541143 Flying Officer (Navigator) **Leslie Frank Stannard** RAFVR of 138 Squadron aged 33 was the tail-gunner aboard Halifax V LL192 on Operation Tablejam 46. The plane had taken off from RAF Tempsford to drop supplies over Denmark. Having dropped the load, at 2.19 a.m. the Halifax was intercepted on its return flight and shot down into the North Sea (more precisely in the Skagerrak Channel) approximately 35 miles (60km) north of Thisted, by Feldwebel Klaus Möller of 12/NJG 3. All on-board the Halifax perished. Living at Harley House, Leighton Buzzard and the son of Frank and Ethel Stannard of Chandler's Ford, Hampshire, he had been headmaster of Beaudesert Boys' School since 1937 and Principal of Leighton Evening Institute. Kviberg cemetery, Göteborg, Sweden Plot 3 Row B Grave 3.

1944 TUESDAY 9 MAY: 40336 Major **Charles Herbert Brent Grotrian** aged 41 of 7th Indian Field Regiment, formerly commander of the Luton Battery of the Bedfordshire Yeomanry, was killed in action in Burma. Husband of Aileen Grotrian of 'Pond Cottage', Heath, and of St John's Wood and son of Sir Herbert Brent and Lady Grotrian of 'The Knolls', Plantation Road, Leighton Buzzard, he had been a barrister-at-law. Taukkyan War Cemetery, Burma, Plot 5 Row H Grave 23.

1944 TUESDAY 16 MAY: 11419377 Private **Albert James Clark** aged 24 of the Beds & Herts was killed in action during the Fourth Battle of Cassino. The son of Harry and Ethel Clark of 61 Stanbridge Road, Leighton Buzzard, he had worked as a plumber for Walter Collett & Son of 51 and 53 Hockliffe Street, Leighton Buzzard. Cassino War Cemetery Plot 12 Row G Grave 19.

1944 MONDAY 29 MAY: 1332583 Flight Sergeant (Navigator/Wireless Operator) **Richard Ernest Gray** RAF of 141 Squadron died aged 23 in an air training crash after taking off from RAF West Raynham, Norfolk. Only he, the observer, and the pilot were in de Havilland Mosquito NF Mk II DD717 when the aircraft spun into the ground at North Farm, Clenchwarton, near King's Lynn. The son of Jesse and Ellen Gray of 65 Summerleys, Eaton Bray, he had worked at Waterlow and Sons Ltd, Dunstable. St Mary's Churchyard, Edlesborough.

1944 MONDAY 29 MAY: 5950866 Private **Sydney Thomas John Hyde** aged 24 of 2nd Battalion Beds & Herts Regiment died of wounds received in military action at Monte Cassino, Italy before 23 May, when news was received at home that he was dangerously ill. He was the son of Thomas and Florence Hyde of 21 East Street, Leighton Buzzard, Cassino War Cemetery Plot 1 Row E Grave 4.

1944 MONDAY 5 JUNE: 6465004 Private **George Dorsett,** aged 25 of the Pioneer Corps, was killed in action 'in the Central Mediterranean theatre of war' after Cassino had been taken by the Allies. The son of Mark and Susan Dorsett of Leighton Road, Linslade, he had worked with his parents on Lawrence Boyd Faulkner's canal boats; after their retirement, he had been employed at the Marley Tile Company in Stanbridge Road, Leighton Buzzard. Cassino War Cemetery Plot 8 Row E Grave 7.

1944 TUESDAY 6 JUNE: CH/X 108711 **Cyril Ernest Stevens** aged 21 of No. 48 RM Commandos was evacuated from Dunkirk and later wounded on the Plain of Catania, Sicily. He was among the 116 naval personnel who died on the beaches during the Normandy Landings. He was the son of Ernest and Nora Stevens, later Read, of 'Ferndale', Dudley Street, Leighton Buzzard and stepson of George Baden Read; he had been apprenticed as a dental mechanic to Herbert Stapleton of 54 Hockliffe Street, Leighton Buzzard. Bayeux War Cemetery, Calvados, France Plot 15 Row B Grave 3.

1944 SUNDAY 25 JUNE: P/JX 131567 Petty Officer **Caruth 'Charles' Main** DSM RN of HM Landing Craft HQ 185 died aged 32 when his vessel was mined during the D-Day operations. Awarded his DSM for his part in the Combined Operations Pilotage Parties in 1943 in Sicily, he had also seen action in Algiers and Malta with COPP 3. The adopted son of Mr and Mrs N. Rogers of Leighton Road, Linslade, he had joined the RN upon leaving school. Portsmouth Naval Memorial, Hampshire Panel 81 Column 2.

1944 MONDAY 26 JUNE: C/JX 408436 Ordinary Seaman **Cyril William Duckett** RN aged 19 was aboard the temporary landing-HQ ship HMS *Nith*, which was co-ordinating shipping into the Mulberry harbour. The funnel and bridgeworks had been painted a vivid red for the merchant ships to locate her easily; she was thus highly visible. At 11.21 on the evening of 24 June 1944, *Nith* was struck on the starboard side by a Mistel. The Mistel was one of the enemy's latest weapons, a composite aircraft of an ME109 and an old JU88 with the crew compartment of the latter replaced and packed with high explosives. The JU88 would then be attached to the underside of its 'parent', which would take it to its target and then release it. The dead were buried at sea at 7 a.m. on 25 June whilst *Nith* was under tow to dry-dock at Cowes. William Duckett died the following day and was presumably among the twenty-six wounded. He was the son of Cyril and Mary Duckett of Burcott, Wing. Southampton (Hollybrook) Cemetery Section M Plot 12 Grave 48.

1944 TUESDAY 27 JUNE: 14205641 Trooper **Frank David Puddephatt** aged 21 a wireless operator with the RAC attached to the Nottinghamshire Yeomanry (Sherwood Rangers) died during Operation Epsom, following the D-Day landings in Normandy, having already served in Egypt and Tunisia. He was the son of George and Jane Puddephatt of 82 Woburn Road, Heath & Reach. Bayeux Memorial Panel 10 Column 2.

1944 THURSDAY 29 JUNE: 14659871 Private **Frank William Pearson** aged 24 of 1st/7th Battalion, The Queen's Royal Regiment (West Surrey) had been a member of the Home Guard before joining the army. His regiment was involved in heavy fighting following the Normandy Landings as Commonwealth forces tried to advance from Bayeux in an encircling movement to the south of Caen. He was the son of Frederick and Edith Pearson of Corner Farm, Edlesborough. Hottot-les-Bagues War Cemetery, Calvados, France Plot IX Row D Grave 3.

1944 SUNDAY 2 JULY: 14556549 **John Henry Hoddinott** aged 19 of 8th (Midland Counties) Battalion, Parachute Regiment of the Army Air Corps, took part in Operation Overlord, the Normandy campaign. After participating in the capture and destruction of bridges over the River Orne, his regiment was mining, laying ambushes and raiding enemy positions in the Bois de Bavent area. He was the son of Henry and Dorothy Hoddinott, of Old Church House, Bierton, Buckinghamshire. Ranville War Cemetery, Calvados, France Plot IIA (2A) Row E Grave 8.

1944 MONDAY 10 JULY: 134500 Major **Robert Frank Gale** MC and Bar aged 25 of C Squadron 3rd County of London Yeomanry (Sharpshooters) of the RAC was hit and fatally wounded during heavy shelling while his regiment was moving across the River Odon to the north of Éterville in Normandy. Husband of Margaret Gale of 144a Wilbury Road, Letchworth and son of Frank and Hilda Gale of Bridge Farm, Stoke Hammond, he had managed a brickyard in Peterborough. Banneville-La-Campagne War Cemetery Plot XI Row E Grave 15

1944 WEDNESDAY 12 JULY: 14695179 Private **Alan Henry Morris** aged 19 of 7th Battalion, Seaforth Highlanders was killed while fighting near Caen in Normandy, probably during or following Operation Jupiter, which had been launched to push back the German forces near the village of Baron-sur-Odon, and to retake Hill 112. Son of Charles and Nellie Morris of Grovebury Road,

Leighton Buzzard, he had been a member of the Leighton Gas Company office staff. Secqueville-En–Bessin War Cemetery, Calvados, France Plot II (2) Row A Grave 14.

1944 MONDAY 17 JULY: 14694981 Private **Kenneth Sidney Woolhead** aged 18 of 1st/17th Battalion, The Royal Warwickshire Regiment died during the Battle for Caen. He was the son of Sydney and Edna Woolhead of New Bradwell, Buckinghamshire and a grandson of Albert and Matilda Woolhead of Bossington Lane, Linslade. Fontenay-le-Pesnel War Cemetery, Tessel, France Plot III Row D Grave 13.

1944 WEDNESDAY 19 JULY: 14288625 Trooper **Kenneth J. Sayell** aged 20 of 'B' Squadron, 5th Royal Tank Regiment of the RAC died in his tank during the liberation of Bourgebus, Manche, according to a member of his household at that time. Another report states that he was seen climbing out of the tank but that his body was never found. 'At 5 p.m. the Cromwells of the 5th Royal Tank Regiment aimed for the ridge of Bourguebus, but they were repelled by *Panzergrenadiere* of the *Leibstandarte* and a Tiger tank of the *Schwere Panzer-Abteilung* 503.' He was the son of Frank and Mildred Sayell, of Rosebery Avenue, Linslade. Bayeux Memorial Panel 8 Column 3.

1944 FRIDAY 28 JULY: 14655895 Fusilier **Philip William Wray** aged 19 of 2nd Battalion Royal Fusiliers (City of London Regiment) was killed by a sniper in Italy after a display of courage and initiative that was highly praised by Major J.H. Draper, his officer. The son of Frederick and Violet Wray of 4 Friday Street, Leighton Buzzard, he had been a member of the 1st Leighton Buzzard Scout Troop and had joined the army upon leaving school. Florence War Cemetery, Italy Plot X (10) Row C Grave 6.

1944 MONDAY 31 JULY: 5947801 Private **Stanley Boyce** aged 32 of 1st Battalion Hampshire Regiment, formerly of the Beds & Herts, died in northern France following the Normandy Landings. His regiment had landed on 6 June at Gold Beach, north of Caen at Courseulles-sur-Mer and by 30 July had fought its way inland, with heavy fighting at St Germain d'Ectot and Launay, to lead the assault towards Villers-Bocage. He was the son of William and Minnie Boyce of 34 Doggett Street, Leighton Buzzard and had worked for the Speight Tile Company, Grovebury Road, Leighton Buzzard. St Charles de Percy War Cemetery, Calvados Plot 15 Row G Grave 7.

1944 MONDAY 31 JULY: 5947238 Trooper **John William 'Jack' Lancaster** aged 33 of 4th Regiment Reconnaissance Corps, RAC, was killed in the aftermath of the taking of Arezzo by the 6th Armoured Division. 'By the death of this son Mr and Mrs Lancaster have lost seven of their nine children.' He was the husband of Alice Lancaster of 9 Baker Street, Leighton Buzzard and son of John and Jane Lancaster of 20 Doggett Street, Leighton Buzzard. After spending some years with the Beds & Herts Regiment in India, he had returned to work for Garside Sand Merchants and was a member of Leighton Hockey Club. Arezzo War Cemetery Plot III Row D Grave 27.

1944 MONDAY 7 AUGUST: 273399 Lieutenant **Alan McQueen Don** aged 20 of the Border Regiment, RAC, seconded to 1st Battalion the King's Own Scottish Borderers, was killed in action in Normandy. His family had moved to Leighton Buzzard from Liverpool during the 1930s. The son of James and Margaret Don of 'Sandy Mount', Plantation Road, Leighton Buzzard, he had trained at Sandhurst after leaving school. Bayeux Memorial Panel 15 Column 1.

1944 TUESDAY 8 AUGUST: EC/2895 Captain **Lionel John Jordan** aged 28 of the Corps of Indian Electrical and Mechanical Engineers died in Arezzo. Twenty-nine others, including Gurkhas, died on the same day and are buried there. On the roll of honour for Beaudesert School, Leighton Buzzard, he is named as Major Jordan of the Royal Army Ordnance Corps. He was the husband of Joyce Jordan of 'Beau Desire' (Beaudesert?), Stanbridge Road, Leighton Buzzard and son of Joseph and Daisy Jordan of Leighton Buzzard. Arezzo War Cemetery, Italy Plot I (1) Row A Grave 21.

1944 Thursday 10 August: 320940 Trooper **William George Hunt** MM aged 26 of the 6th (Airborne) Armoured Reconnaissance Regiment was in a house at the Divisional Rest Camp in Ouistreham, Normandy when an enemy aircraft crashed into the building at around 10.45 p.m., killing him and another member of HQ Squadron. He was the husband of Sheila Hunt of West Brompton and the son of Henry and Ada Hunt of Northall. Ranville War Cemetery Plot 5a Row E Grave 9.

1944 FRIDAY 11 AUGUST: D/JX 444764 Able Seaman **Charles Dennis George Eggleton** RN died aged 20 while serving on HMS *Albatross*, a seaplane carrier that was torpedoed during the Normandy deployment. The husband of Hilda Eggleton of Leighton Buzzard and son of Charles and Edith Eggleton of 20 Broomhills Road, Leighton Buzzard, he had been employed as a bricklayer by Albert Sear of Hartwell Crescent. Plymouth Naval Memorial, Devon Panel 86 Column 2.

1944 SUNDAY 13 AUGUST: 172481 Flying Officer Pilot **Reginald Sidney Hancox** RAFVR aged 29 of 12 Squadron of No. 1 Group Bomber Command was based in Wickenby and had flown on many operations during June and July 1944. On Saturday 12 August, he was aboard Avro Lancaster PB247, which took off at 9.17 p.m. on a mission to Braunschweig but failed to return, crashing in Eickenrode, Lower Saxony. He was the husband of Winifred Hancox of Toddington and son of John and Winifred Hancox. Hanover War Cemetery Plot 8 Row B Collective Grave 12-17.

1944 THURSDAY 7 SEPTEMBER: 5957712 Fusilier **Donald Mead** aged 28 of 6th Battalion The Royal Scots Fusiliers died during the liberation of Derik in Belgium, which had begun on 5 September. It ended on 7 September after very heavy shelling and bombing from the neighbouring village of Zwevegem, which had already been freed by the Allies. The four Commonwealth graves in the cemetery there are all of soldiers from Scottish regiments. He was the husband of Emily Mead of 116 Kuella Road, Welwyn Garden City, son of Lily and stepson of Benjamin Gray of 'The Stables', Edlesborough and had assisted in his stepfather's business before the outbreak of war. Deerlijk Communal Cemetery Grave 2.

1944 TUESDAY 12 SEPTEMBER: On 4 September 1944, convoy HI-72 sailed from Singapore, two of these transport ships, *Rakuyo Maru* and *Kachidoki Maru*, carrying British POWs from the Burma–Thailand railway, which had by then been completed. On 12 September the convoy was attacked by US submarines and both ships were hit. *Rakuyo* was torpedoed by the US submarine *Sealion* at around 5 a.m. and sank east of Hainan Island, off China, Lat 18.0N Lon 114.0E. *Kachidoki* was torpedoed by the US submarine *Pampanito* at 10.40 p.m. and sank north-east of Hainan Island. When the US submarines realised that they had sunk ships containing POWs, they searched for survivors, picking up sixty-three from *Rakuyo*, four of whom died soon after. Of Rakuyo's prisoners, 1,159 were lost to the sea and the after effects of being in the water for up to four days. The Japanese rescued some of the POWs from these two ships and continued their journey to Japan on the *Kibitsu Maru*.

These six local men missing at sea from the *Rakuyo* had been taken prisoner with their regiment on 14/15 February 1942 in Singapore:

871753 Lance Sergeant **Ronald Henry Dennison** aged 24 of 512 Battery the Bedfordshire Yeomanry, son of Henry and Alice Dennison of 12 Suncote Avenue, Dunstable. Singapore Memorial, Singapore Column 36.

925369 Bombardier **Stanley Frederick Hart** aged 24 of 419 Battery the Bedfordshire Yeomanry, son of Henry and Nellie Hart of 94 Soulbury Road, Linslade. Singapore Memorial, Singapore Column 36.

917289 Gunner **Bernard Frederick Russell Hedges** aged 24 of 419 Battery the Bedfordshire Yeomanry, a wireless operator, son of Frederick and Florence Hedges of 14 Mentmore Road, Linslade; he had worked in the machine room at Waterlow & Sons Ltd, Dunstable. Singapore Memorial, Singapore Column 37.

917317 Lance Sergeant **Dennis Alfred Horne** aged 26 of 512 Battery the Bedfordshire Yeomanry, son of Oscar and Adeline Horne of 120 Wing Road, Linslade; he had been employed by Messrs John Dickinson of Apsley Mills and was a member of Leighton Town Cricket Club. Singapore Memorial, Singapore Column 36.

913912 Lance Sergeant **Harold John Law** aged 28 of 419 Battery the Bedfordshire Yeomanry, husband of Dorothy Law of 8 Kent Road, Luton, formerly of 47 South Street, Leighton Buzzard, and son of George Frederick and Mary Law; he had worked at Vauxhall Motors Ltd, Luton. Singapore Memorial, Singapore Column 36.

9233011 Lance Bombardier **Harold Trevor Williams** aged 23 of 512 Battery the Bedfordshire, son of Gwilym and Eliza Williams of 'Haughmond', 95 Wing Road, Linslade. He had been employed by Messrs John Dickinson of Apsley Mills and was a member of Leighton Town Cricket Club. Singapore Memorial, Singapore Column 37.

These three soldiers were reported missing at sea on or shortly after the same date from the *Kachidoki Maru*:

918241 Gunner **Kenneth Bailey** aged 24 of 419 Battery the Bedfordshire Yeomanry was reported by a survivor to have fallen off a raft, being too weak to save himself. The son of Reuben and Elizabeth Bailey of 13 Ashburnham Crescent, Linslade, he had been employed by E. Albertini and Company (England) Ltd of Morgan Works, Leighton Road Linslade, making felt hoods for Luton's hat industry. Singapore Memorial Column 37.

5953352 Corporal **Cyril Green** aged 28 of 5th Battalion Beds & Herts was the husband of Eileen Green of 17 Brook Street, Tring, Hertfordshire and the son of Mrs W.G. Sutton of Stanbridge Crossing, Leighton Buzzard Singapore Memorial Column 61.

873278 Lance Sergeant **Donald Edwin Horne** aged 25 of 419 Battery the Bedfordshire Yeomanry, was the son of Sydney and Edith Horne of 138 Luton Rd, Dunstable and had been employed in the Luton Rural District Council Surveyor's Depot. Singapore Memorial Column 36.

1944 SUNDAY 17 SEPTEMBER: 2130621 Sapper **Frederick George Badcock** aged 35 of 87th Assault Squadron, RE died of wounds received during the attack on Boulogne. Husband of Mary Badcock of High Street, Leighton Buzzard and son of Albert and Eliza Badcock, he had worked for Gossard and had been a member of the choir at St Andrew's church. Calais Canadian War Cemetery, Leubringhen, Pas de Calais Plot 3 Row B Grave 1.

1944 WEDNESDAY 20 SEPTEMBER: Sub-Lieutenant (A) **Peter William Pidgley** RNVR of HMS *Macaw* (Royal Naval Air Station Bootle) died aged 19 during a training course with the 9th Pilot Advanced Flying Unit of the Fleet Air Arm. *Macaw* was a Fleet Air Arm station in Bootle, Cumberland, where new pilots assembled after their preliminary flying training in Canada and North America on their way to advanced naval flying units in Britain. He was the son of George and Louisa Pidgley of 'Vulpera', 8 Aylesbury Road, Wing and had joined the Fleet Air Arm immediately after leaving school. All Saints' Churchyard, Wing Row 22 Grave 10.

1944 THURSDAY 21 SEPTEMBER: 5955225 Private **Geoffrey Chorlton** aged 24 of 5th Battalion Beds & Herts Regiment was lost at sea. On 21 September 1944 the *Hofuku Maru* sailed with Convoy MATA-27 for Takao in Formosa. She was attacked by an American aircraft carrier about 80 miles north of Corregidor; the aircraft carrier's planes sank the whole fleet, including the *Hofuku*, not knowing that she was carrying 1,289 prisoners from Manila to Japan, 1,047 of whom were lost. He was the son of Eric and Ethel Chorlton of Dunstable. Singapore Memorial Column 62.

1944 FRIDAY 22 SEPTEMBER: 6898297 Private **Paul Silwood Humphries** aged 25 of 1st/6th Battalion Welch Regiment died shortly after the liberation of Nederweert, Holland by British troops on 21 September 1944. The husband of Doreen Humphries of Linslade and the son of James and Frances Humphries of Linslade, latterly of Wembley, he had been employed by the General Electric Company, London, Nederweert War Cemetery, Netherlands Plot III (3) Row E Grave 1.

1944 SATURDAY 23 SEPTEMBER: 5834205 Trooper **Frederick John Stone** aged 31 RAC attached to 142nd (7th Battalion, The Suffolk Regiment) was in Italy, defending Coriano Ridge. On the night of 12 September the Eighth Army had reopened its attack on the ridge. This attack successfully took the ridge, but marked the beginning of a week of the heaviest fighting experienced since Cassino in May, with daily losses for the Eighth Army between 100 and 200. He was the son of Joseph and Annie Stone of Heath and Reach. Coriano Ridge War Cemetery Plot XIX Row C Grave 6.

1944 SUNDAY 15 OCTOBER: 1063511 Gunner **Reginald William Woodland** aged 37 of 76th Field Artillery RA is likely to have died during the battles for Overloon and Venray in the south-east of Holland, when heavy Allied losses were suffered. He was the son of Alfred and Lily Woodland of Fosse House, Clandown, Radstock, Somerset. While working for three years in the stables at Ascott House, he had lived above the stables; he later joined the Leighton Buzzard postal staff and moved to 21 Beaudesert, Leighton Buzzard. Overloon War Cemetery, Noord-Brabant, Netherlands Plot II (2) Row A Grave 13.

1944 SUNDAY 29 OCTOBER: 1872415 Sergeant (Wireless Operator/Air Gunner) **Henry George Smith** RAFVR aged 20 of 207 Squadron based at Spilsby, Lincolnshire, was killed along with the seven other members of the crew of Avro Lancaster LM271 EM-L, which was taking part in a raid on a U-boat bunker codenamed Bruno in the Norwegian port of Bergen. Thicker cloud than anticipated forced them lower. The plane crashed in the target area and only the bodies of the two rear-gunners were found initially. Sergeant Smith's death was not officially confirmed until July 1945. The plane, one of 237 Lancasters to attack the bunker that evening and one of the eight lost, was hit by flak. The son of Robert and Ada Smith of 8 The Square, Southcourt, Linslade, he had been employed as chief clerk by Thornley and Boutwood, solicitors, of 13 High Street, Leighton Buzzard. Bergen (Mollendal) Church Cemetery Plot N Row D Collective Grave 2.

1944 SATURDAY 18 NOVEMBER: 5954936 Lance Corporal **Frederick Linney** aged 24 of 2nd Battalion The Royal Scots Fusiliers was killed instantly by a shell landing in a trench during the breaking of the Gothic line along the summits of the northern part of the Apennine Mountains in Italy. He was the adopted son of Samuel and Alice Linney of 38 Bedford Street, Leighton Buzzard. Faenza War Cemetery Plot VI Row C Grave 16.

1944 TUESDAY 19 DECEMBER: 595401 Private **Herbert George Billingham** aged 29 of 4th Battalion the Buffs (Royal East Kent Regiment) was killed on the day he was reported missing in Athens. The husband of Rhoda Billingham of 58 Bassett Road, Leighton Buzzard, he had worked on the Grand Union Canal in his youth and later for the Marley Tile Company in Stanbridge Road, Leighton Buzzard. Phaleron War Cemetery Plot 18 Row E Grave 9.

1944 WEDNESDAY 27 DECEMBER: 923006 Gunner **Norman Mayo Missenden** aged 30 of 512 Battery the Bedfordshire Yeomanry was captured on 14/15 February 1942 in Singapore. He died at Sendai 9 POW Camp in Osaka, the cause of his death being beriberi and an ulcerated left leg. He was the son of Algernon and Florence Hannah Missenden of The Hollies, and later The Gables, Linslade, and had worked for his father, a corn merchant. Yokohama War Cemetery, British Section Plot L Row B Grave 10.

1945 MONDAY 1 JANUARY: 104559 Flight Lieutenant **Alfred Stewart Smith** RAFVR aged 35 of 193 Squadron was flying a single-seater Hawker Typhoon RB218 to Holland with a railway as his target. At 4.15 a.m. his plane was hit by German flak, caught fire and plunged down behind a shipyard by the River Waal at Zaltbommel. He had lived with his wife at 2 Stewkley Road, Wing. Zaltbommel General Cemetery, Netherlands Row A Grave 4.

1945 SUNDAY 21 JANUARY: 1088439 Bombardier **Frederick George Jeffs** aged 33 of 160th Field Regiment RA was based in Burma. Bombardier Jeffs was killed in an assault on Ramree Island off the coast of Burma. He was the husband of Rose Jeffs of 1 Waterloo Road, Linslade, and son of William and Emily Jeffs of 20 Rosebery Avenue, Linslade. Rangoon Memorial, Myanmar Face 2.

1945 MONDAY 29 JANUARY: 156318 Captain **John Charles Franklin** aged 28 of the Beds & Herts Regiment, attached to 1/5th Battalion The Queen's Royal Regiment (West Surrey), died in Holland during the campaign to force the Germans back across the River Maas (Meuse). He had lived at 'Woodlands', Plantation Road, and was the son of Charles and Maud Franklin of Harpenden, Hertfordshire, formerly Coal and Coke Merchants, 31 High Street, Leighton Buzzard. Nederweert War Cemetery Plot III Row D Grave 11.

1945 WEDNESDAY 31 JANUARY: EX. 3192 Corporal **William Joseph Dell** aged 25 of No. 42 RM Commando Unit died in Burma. During the third Arakan campaign his unit, landed on the peninsula on 12 January and were defending Hill 170 by the 30th. A day later the Japanese mounted a suicide attack, destroying two of the three Sherman tanks by climbing on them and exploding charges fixed to the ends of bamboo poles. Attack was followed by counter-attack, an estimated 700 Japanese artillery shells landing on the hill during the continuous fighting on that last day of the battle. He was the son of Walter and Bithia Dell of 60 Mill Road, Leighton Buzzard, and had worked for Charles Mares, outfitters, of Luton, having been apprenticed to Leonard Croxford of North Street, Leighton Buzzard. Taukkyan War Cemetery, Myanmar Plot 11 Row D Grave 4.

1945 FRIDAY 2 FEBRUARY: 1553022 Lance Sergeant **Frederick John Foster** aged 28 of 67th (Suffolk) Medium Regiment, RA was killed by a land mine. In 2012 a Dutch woman contacted everyone named Foster in the Chesham area in an attempt to trace any of his relatives; 'as part of a local tradition' her grandmother had always visited and tended his grave and she wished to contact his family to ask if she could take over from her now very elderly relative. Husband of Dorothy Foster of Chesham, Buckinghamshire and son of William and Sarah Foster of 29 Broomhills Road, Leighton Buzzard, he had worked for the London Brick Company Ltd of Mile Tree Road, Leighton Buzzard. Brunssum War Cemetery, Limburg, Netherlands Plot VI Grave 307.

1945 SUNDAY 4 FEBRUARY: 1553045 Private **William Henry Lee** aged 29 of 1st Battalion the Lincolnshire Regiment was killed in action in Burma. He was the husband of Lilian Lee of Dunstable and son of Albert Thomas and Minnie Lee of 92 Soulbury Road, Linslade, and had worked for the Stonehenge Brick Company, Leighton Buzzard. Taukkyan War Cemetery, Myanmar Plot 10 Row D Grave 4.

1945 WEDNESDAY 7 FEBRUARY: 5945090 Lance Sergeant **Sidney Cyril Flemming** aged 42 of 5th Battalion Beds & Herts was captured with his regiment on 14/15 February 1942 in Singapore. He was killed in a US air raid on Formosa, probably while in Heito Camp. The wooden huts offered little protection even from shrapnel. He was the husband of Hilda Flemming of 34 Lake Street, Leighton Buzzard, and son of Thomas and Emma Flemming. Sai Wan War Cemetery, Hong Kong Plot 7 Row F Grave 12.

1945 SUNDAY 11 FEBRUARY: 14991981 Private **Kenneth William Stanbridge** aged 19 of 2nd Battalion Argyll & Sutherland Highlanders died near Nijmegen in Holland during the Battle for the Rhineland. Son of Robert and Blanche Stanbridge of Stanbridge Road, Leighton Buzzard, he had worked for W.E. Wallace and Son, nurserymen, of Eaton Bray. Jonkerbos War Cemetery, Nijmegen Plot 21 Row G Grave 5.

1945 THURSDAY 15 FEBRUARY: 1615776 Flight Sergeant **John Joseph O'Brien Heady** RAFVR aged 22 of 153 Squadron based at Scampton died when his plane was shot down by a JU88 over eastern Germany. Dresden, the capital of Saxony, was a major rail transportation and communication centre with over a hundred factories, a prime military and industrial target. Operation Thunderclap had begun with a major Bomber Command attack on 13 February; on the night of the 14th, 153 Squadron supplied nine aircraft to join a raid on Chemnitz, 80km south-west of Dresden. One of these, Avro Lancaster NN803

(P4-O), had taken off from RAF Scampton at 8.05 p.m. but crashed, burning fiercely, at 12.44 a.m. on 15 February between the villages of Arnsfeld and Grumbach, having being targeted by Oberfeldwebel Ludwig Schmidt piloting a JU88G-6 from 2./NJG6. The crew were all killed. Flight Sergeant Heady was on-board as flight engineer; he had trained as a pilot but was one of many re-mustered either as flight engineer or glider pilot, both of which were in short supply, whereas there was a glut of available pilots at that time. Three of the crew are buried in the Berlin 1939–45 War Cemetery but the other four men have no known graves. He was the son of Ernest and Bridget Heady of 58 Southcourt Avenue, Linslade, and had worked in the Woburn Sands branch of the Westminster Bank. Runnymede Memorial, Surrey Panel 271.

1945 FRIDAY 16 FEBRUARY: 78086 Flight Lieutenant (Navigator) **Kenneth Douglas Faulkner** DFC RAF aged 38 of 1674 HCU, 17 Group RAF Coastal Command was killed in a flying accident. No. 9 Liberator LL Course had commenced training earlier in the month and his plane, Liberator GR.VI – serial number EV954 – flew into high ground during a night-time operational flying exercise near Collinvale, ¾ mile north-west of Ballyclaire, County Antrim. His DFC was awarded while he was with 15 Squadron, navigation officer and bombing leader with Bomber Command, taking part in raids over Cologne and Essen in 1942. The son of Lawrence and Annie Faulkner of 'The Vale', Stoke Road, Linslade, he had worked for the Percival Aircraft Company at Luton. St Mary's Churchyard, Old Linslade Section a Grave 59.

Flight Lieutenant Kenneth Douglas Faulkner's gravestone at St Mary's church, Old Linslade.

1945 SATURDAY 10 MARCH: 930000 Gunner **Rudolphe Gaston Bastin** aged 24 of the Bedfordshire Yeomanry was captured with his regiment on 14/15 February 1942 in Singapore. He was in both Taichu and Kinkaseki camps on Formosa. In the former, the main work was the excavation of a flood diversion channel in the river valley that ran adjacent to the camp. The latter was the base for POWs who were to be used as slave labour in Japan's largest copper mine, and it was here that Rudolphe died. The son of Gaston and Beatrice Bastin and the adopted nephew of Elsie Pratt of 'Rendlesham', Bedford Street, Leighton Buzzard, he had worked for Oliver Ryland Ltd and had belonged to the 1st Leighton Buzzard Scout Troop. Sai Wan War Cemetery, Hong Kong Plot 5 Row E Grave 3.

1945 MONDAY 9 APRIL: 330715 Captain **Richard William Butler** aged 30, an Animal Transport Officer of the Beds & Herts Regiment, attached to 9th Battalion Border Regiment, died shortly after the Allies had captured Mandalay and were moving south through the Irrawaddy Valley towards Rangoon. It is likely that Captain Butler died during the battle for Pyawbwe. He was the son of William and Margaret Butler of 'Timberedge', Hillside Road, Leighton Buzzard, and had bought and run a farm in Lidlington before selling his land and joining the services. Taukkyan War Cemetery, Myanmar Plot 20 Row H Grave 18.

1945 Monday 9 April: Adjutant **Ernest Robert Gaskin** aged 35 of the Salvation Army was transferred at his own request to the Salvation Army 'Red Shield' Mobile Canteens at various camps in England and later in Ireland and by June 1944 was 'operating a Welfare Centre' in Italy. He was killed while running a canteen on the docks at Bari. Son of Thomas and Elizabeth Gaskin of 'Fairview', Queen Street, Leighton Buzzard, he had at one time been employed by Griffin Brothers as manager at their shop in North Street, Leighton Buzzard. Bari War Cemetery, Italy Plot 9 Row D Grave 9.

1945 SATURDAY 14 APRIL: 6094808 Sergeant **William James Hack** aged 26 of 56th Regiment Reconnaissance Corps, RAC, died of an accidental gunshot wound during the Battle of Argenta Gap in northern Italy. He was the husband of Hilda Mary Hack of 130 Soulbury Road, Linslade, and son of Joseph and Edith Hack of Heath and Reach. Faenza War Cemetery Plot II (2) Row C Grave 19.

1945 FRIDAY 29 JUNE: Pilot Officer (1st Class) **Ernest Arthur David Kempster** aged 44 of Air Transport Auxiliary, Air Movements Flight, had always courted danger. Known locally as Snuffy and more widely as 'Smiling Jim' Kempster, he was England's first speedway captain and the man who led his country in two of the first three speedway tests against Australia in 1930. He gave up the sport after 'a bad crash' in the mid-1930s, but appears to have continued to enjoy taking risks: during the 1930s a flight with him from an airfield on Scott's Field, Billington Road would cost the passenger 4s (20p); this would be in the region of £11 in 2013. 'Jim' died on or shortly after Friday 29 June 1945, just two weeks after being commended for valuable service in the air in the King's

OGDEN'S CIGARET

'Jim' Kempster's life was filled with excitement: speedway had only arrived in England in 1928 and he was immediately a star, captaining the England team in 1930. He was also an experienced pilot and when war broke out he willingly took the same risks as the RAF fliers joining the Air Transport Auxiliary.

birthday honours list. He was ferrying an Anson DG916 from Le Bouget in France to Pilsen in Bohemia: bad weather caused poor visibility and his plane appears to have struck overhead cables strung across the Rhine near Koblenz and crashed into the river. He was the husband of Hylda Kempster of Peddars Close, Stanbridge, and the son of Walter and Florence Kempster of 'Fern Villa', Clarence Road, Leighton Buzzard, and had achieved international fame as captain of the All England Dirt Track racing team, representing England in Australia and New Zealand. Rheinberg War Cemetery, Nordrhein Westfalen, Germany Plot 3 Row K Grave 22.

1945 FRIDAY 10 AUGUST: T/14603557 Sergeant **Douglas Francis Paxton** RASC attached to 44 General Transport Company, East African Army Service Corps died aged 20 in East Africa. During the Second World War, Moshi was the HQ of Southern Area with a major-general's command and a concentration centre for troops destined for India and Burma. He was the son of Doris Paxton of Wing. Moshi Cemetery, Tanzania New Section 2 Row B Grave 1.

1945 SATURDAY 18 AUGUST: Flying Officer **Stanley John Abbott** RAF aged 38 had relinquished his commission 'on cessation of duty' on 5 December

1944, possibly because of ill health. His death at The Royal Victoria Hospital, Belfast, was due to an illness contracted on military service in the Middle East. The son of Sydney and Maria Abbott of 'Hillmount', 29 Stoke Road, Linslade, he had gained qualifications with the Customs and Excise Department and had been District Scoutmaster of Linslade and District Boy Scouts Association. St Mary's Churchyard, Old Linslade Plot C Grave 34.

The end of the war did not mean the end of the deaths; many were still stationed at home or abroad and some were making their way home.

1945 MONDAY 17 SEPTEMBER: EC/14011 Major **Dennis Page** aged 23 of the Royal Indian Army Service Corps died in India. He was the son of William and Alice Page of Oxley's Farm, Aylesbury Road, Wing, and latterly of Tring. Delhi War Cemetery Plot 1 Row D Grave 11.

1945 SATURDAY 29 SEPTEMBER: T/5957728 Driver **Cyril James West** RASC died aged 30 in Avro Lancaster MKI PD343. This was on a transit flight from Italy, carrying army servicemen back to the UK during an Operation 'Dodge' sortie, and the plane was reported lost without trace, possibly after entering cumulo-nimbus cloud. He was the husband of Gladys West and son of Harry and Kate West of Leighton Buzzard. Brookwood Memorial, Surrey Panel 17 Column 2.

1945 WEDNESDAY 14 NOVEMBER: 5959076 Craftsman **William Patterson Jardine** aged 30 of 539 Workshop Company Royal Electrical and Mechanical Engineers died in hospital of septicaemia 'as a result of typhoid and related diseases' at Tel El Kebir Garrison in Egypt, 110km north-north-east of Cairo and 75km south of Port Said on the edge of the desert. The husband of Grace Jardine of 107 Stanbridge Road, Leighton Buzzard, and son of William and Jeannie Jardine of Leighton Buzzard, he had been employed at Vauxhall Motors Ltd, Luton. Tel El Kebir War Memorial Cemetery, Egypt Plot 6 Row O Grave 5.

1946 MONDAY 4 MARCH: CH/X113236 Royal Marine **Donald Frank Griffin** aged 21 died 'of wounds' according to a report in a Bedfordshire newspaper. He was stationed in Singapore at shore station HMS *Simbang*, a Royal Naval Air Station. He was the son of Herbert and Dora Griffin of Linslade and had been apprenticed at Vauxhall Motors Ltd. Kranji War Cemetery, Singapore Plot 35 Row C Grave 9.

1946 MONDAY 21 OCTOBER: 1616906 Gunner **Stanley Bray** RA died aged 32 in the Aylesbury area. He had been posted to an anti-aircraft gun placement in Ireland; while in service, he became ill with a virus from which he died. The husband of Peggy Bray of Leighton Buzzard and son of Adam and Elizabeth Bray, he had been a bandsman and songster in the Salvation Army. Leighton Buzzard Cemetery Section HH Grave 63.

1947 FRIDAY 29 AUGUST: 5944425 Driver **James Edward Thomas Kempster** RASC, born in Leighton Buzzard and formerly of the Beds & Herts Regiment, died in London – his death aged 41 was registered in Stepney registration district. He was the husband of Mary Kempster of Stepney, and son of

Army Form B2089 – a soldier's will. This one is dated 26 October 1940 and signed by Tom Agutter Skevington, a private in the Royal Army Ordnance Corps, appointing his father, Tom, as his executor and leaving everything to his wife Edith of 46 George Street, Leighton Buzzard. The will was not needed: he survived the war and took over the family shoe shop in the High Street.

Thomas and Sarah Kempster of 6 Barrow Path, Leighton Buzzard. East London Cemetery, Grange Road, Plaistow Grave 16374; Screen Wall Panel 8.

1947 WEDNESDAY 29 OCTOBER: T/237999 Private **John Henry Syratt** RASC died aged 39 in the area where he lived. He was the husband of Kathleen Syratt of Kingsbury, Middlesex, and the son of James and Jane Syratt of 3 Long Row, Burcott, Wing. All Saints' Churchyard, Wing Row 24 Grave 9.

PART FIVE

APPENDICES

TIMELINE 1938-45

1938

JANUARY

The Air Raid Precautions Act comes into force, compelling local authorities to set up and train ARP forces (later Civil Defence) and expand their police and fire services. ARP Branches include wardens, rescue personnel, first aid parties, ambulance, messenger and decontamination squads. In all, about 150,000 people are eventually trained in ARP duties.

The Women's Voluntary Service for Air Raid Precautions is established under the leadership of Stella Isaacs, the Dowager Marchioness of Reading, as a support unit for the ARP. Its terms of work are set out in the Air Raid Precautions Act and in the pamphlet 'What You Can Do'. By the end of the year, 32,000 women have enrolled.

SEPTEMBER
The Ministry of Health contacts Bedfordshire County Council about potential accommodation for evacuees.

OCTOBER
The Munich agreement is signed. Prime Minister Neville Chamberlain returns home saying 'I have in my hand a piece of paper' and claiming 'peace for our time'. He is seen as a heroic figure.

1939

FEBRUARY

Despite the prime minister's assurances, ARP continue. The first Anderson shelters, made from corrugated iron sheets, are delivered to householders in

Islington. Eventually, 1.5 million Andersons are brought into use. A replica can be seen at Milton Keynes Museum.

MARCH

German troops march into Czechoslovakia, following 'a quarrel in a far away country between people of whom we know nothing', according to Chamberlain.

JULY

Mr H.E. Baines, Director of Education at Shire Hall, Bedford, sends a letter to Leighton Buzzard schools saying that it is necessary to start practising for air raids at once. 'Every effort must be made to avoid complacency on the one hand, and panic on the other.'

The Home Office asks Leighton Buzzard and Linslade to be prepared to find homes for 1,806 evacuees should an emergency arise.

The Civil Defence Act specifies that factories, mines and commercial buildings employing more than thirty persons are required to have an ARP scheme, meaning that they have to organise, train and equip 10 per cent of their staff to form first aid parties, first aid posts, fire parties, rescue squads, and a works' fire brigade equipped with major appliances. In smaller establishments the total numbers involved in ARP can be reduced by the formation of general utility squads trained to do several jobs.

A First Aid Post and Ambulance Depot is set up at the Public Assistance Institution (the old Workhouse) in Grovebury Road, Leighton Buzzard, together with a Rescue Party Depot and a Decontamination Squad. A Mobile First Aid Unit is located at 3 Leighton Street in Woburn, and First Aid Points at the Methodist chapel, Eaton Bray, the schoolroom in the Methodist chapel at Eggington, the Village Barn at Heath and Reach, the Men's Club at Hockliffe, 76 North Street in Leighton Buzzard and the Methodist chapel schoolroom at Stanbridge.

AUGUST

The National Air Raid Precautions Animals Committee is formed. It drafts a notice entitled 'Advice to Animal Owners'. The pamphlet says: 'If at all possible, send or take your household animals into the country in advance of an emergency.' It concludes: 'If you cannot place them in the care of neighbours, it really is kindest to have them destroyed.' In subsequent weeks, 750,000 animals are put down.

SEPTEMBER

Following the order to evacuate, given in a radio broadcast on 31 August in anticipation of the outbreak of war, 800 children arrive at Leighton Buzzard the very next day on one of the first dedicated evacuee trains out of London.

On 1 September, German troops march into Poland. War is now inevitable.

Broadcasting on 3 September from the Cabinet Rooms in 10 Downing Street, Chamberlain states:

> This morning the British Ambassador in Berlin handed the German Government a final note, stating that, unless we heard from them by 11 o'clock that they were prepared at once to withdraw their troops from Poland, a state of war would exist between us. I have to tell you now that no such undertaking has been received and that consequently this country is at war with Germany … It is evil things that we will be fighting against – brute force, bad faith, injustice, oppression and persecution – and against them I am certain that the right will prevail.

National registration identity cards, petrol rationing and the blackout are introduced.

The government urges people to cut down on car journeys as much as possible, issuing posters asking 'is your journey really necessary?'

OCTOBER

The first bombs land in Britain, at Hoy in the Orkneys (17th).

Coal, gas and electricity are rationed.

The government announces that all men between 18 and 41 not in reserved occupations can be called up to the armed services.

The Defence of the Realm Regulations require shops to close at 6 p.m. in the week and at 7.30 p.m. on Saturdays, the only exceptions being tobacconists, confectioners and newsagents.

1940

JANUARY

Food rationing is introduced at a rate of 4oz (100g) butter, 12oz sugar, 4oz bacon and ham per week. It becomes a criminal offence to waste food.

As the war continues, cheese, margarine, cooking fat, milk, preserves, tea, eggs and sweets are also rationed. All householders have to register with retailers to obtain these items.

Under the Anti-Profiteering Act it becomes an offence to sell specified goods above a certain price.

MARCH

Meat is rationed to 1*s* and 10*d* worth per week, except for the under sixes, who receive 11*d* worth.

Two classes at St Barnabas Infants School in Linslade are amalgamated due to the shortage of coke for fires. Neither classroom can be used because of the cold and all the children are taught in the hall.

A three-hour series of exercises is carried out in Leighton Buzzard to test the efficiency of all sections of the local ARP personnel. Five 'bombs' are dealt with.

APRIL

Lord Woolton becomes Minister of Food, determining that rationing should be used to 'stamp out the diseases that arise from malnutrition, especially those amongst children, such as rickets'. He introduces a national milk scheme for providing a pint of milk a day to children under 5, expectant and nursing mothers and 'invalids with a doctor's certificate'. Only after these priority needs are met can the rest be divided up for general sale.

MAY

Chamberlain resigns the premiership on 10 May 1940, after the Allies are forced to retreat from Norway. He says he believes a government supported by all parties is essential and that the Labour and Liberal parties would not join a government headed by him. He is succeeded by Winston Churchill.

Operation Dynamo, the 'Mosquito Armada', rescues 338,000 men from the beaches of Dunkirk (27 May–4 June)

JUNE

Churchill sets out the challenge: 'Let us brace ourselves to our duties, and so bear ourselves, that if the British Empire and the Commonwealth last for a thousand years, men will still say "this was their finest hour."'

The Ministry of Food issues a bulletin with the advice: 'We should eat more Potatoes. They give energy and protection against illness. They are home-grown. More Potatoes mean less bread and fewer ships to bring wheat from overseas.'

The ringing of church and chapel bells is banned nationally, so that military authorities can use them to warn of the dropping of enemy parachute troops.

Fifty-six iron posts in the High Street of Leighton Buzzard are removed to provide nearly 6 tons to the pile of scrap metal required in the words of the time 'to keep German cattle – or should it be swine – off the nation's pavements' (LBO, 11 June 1940). This proves a difficult task as each post weighs more than 2cwt (100kg) and is solidly embedded in concrete.

JULY

An intense air battle between the Germans and the British takes place over Britain's airspace from July 1940 to May 1941, the heaviest fighting from July to October 1940. It results in a decisive British victory. 'Never in the field of human conflict was so much owed by so many to so few' is Churchill's famous quote about the Battle of Britain.

AUGUST

The first air raid on London takes place, when an off-course flight of Heinkel IIIs drops its bombs on the capital instead of the intended targets of aircraft factories and oil refineries in Kent.

SEPTEMBER

Air raid warnings on the 3rd from 10.50 a.m. to 11.20 a.m. cause children from St Barnabas Infants School to be taken into the church vestry, designated as the school shelter. On the 9th only thirty children out of fifty-eight attend school in the morning, due to repeated air raid warnings in the night.

Two high-explosive bombs fall at the southern end of Woburn village (25th) and one at Potsgrove. The latter fails to explode and is dealt with the next day. Three high-explosive bombs and several incendiaries are dropped near RAF Stanbridge in a direct line from Leighton Buzzard water tower to Eggington village, causing damage to The Elms in Eggington and to Briggington Nurseries at Eggington but no casualties.

OCTOBER

Five high-explosive bombs are dropped near Sand House on the 3rd in open ground alongside the A5. One dropped on Potsgrove fails to explode. Ten days later three high-explosive bombs are dropped near Hockliffe, 200yd (180m) along the Woburn Road, by a low-flying German 'plane passing from north to south'. An unexploded bomb at Potsgrove, probably from the same stick, is dealt with.

On the 14th yet another unexploded bomb is found at Potsgrove, 100yd north of the junction of Sheep Lane and the A5, near Wood Cottage.

Some 240 incendiary bombs fall in the Milton Bryan/Battlesden area on the 18th, plus three high-explosive (HE) bombs in fields at Castle Farm, Battlesden. The following day, twenty-five incendiary bombs fall on Heath and Reach and an HE bomb on a field at Briggington. Three cyclists are injured on Hockliffe Road.

Oil bombs are dropped at Kiteley's Farm and Claridge's Farm and an HE bomb on Inwards' Farm (21st); five bombs land along Stanbridge Road between the Marley Tile Company and the George and Dragon public house, but fail to explode. The following day a high-explosive bomb falls on the chicken run at Pinklehill and one at the bottom of Birds Hill.

1941

JANUARY

The first civilian fatality in the Leighton Buzzard area occurs on the 21st, when an incendiary bomb drops on Heath and Reach, a metal fragment killing a young woman on the road.

JUNE

The BBC introduces *Worker's Playtime* to provide concerts for factories at lunchtime.

AUGUST

The NFS is created, incorporating local brigades and the volunteer Auxiliary Fire Service force, called 'heroes with grimy faces' by Churchill. At its height the NFS reaches 350,000 members.

SEPTEMBER

St Barnabas Infants School is in the grip of a whooping cough epidemic, which lasts until the end of November, causing continuous low attendance.

NOVEMBER

Continued fear of invasion leads to Home Defence Committees organising rest and emergency feeding centres, one of which is St Barnabas church.

DECEMBER

Japanese aircraft attack the US Pacific Fleet at Pearl Harbor (7th). The following day the Japanese invade Malaya.

1942

JANUARY

A national waste paper campaign is initiated to provide war factories with essential materials. Leighton Buzzard collects 39 tons of paper, nearly 11lbs (4kg) per head.
    A combined military and civil exercise is carried out in Leighton Buzzard

and Linslade to test the effectiveness of 'E' Company of the Home Guard in combating 500 airborne troops (played by regulars) who supposedly had landed at Battlesden and were advancing on Leighton Buzzard. The attackers overwhelm the Home Guard by 'weight of numbers', but valuable lessons are said to be learnt.

The 5th Battalion of the Bedfordshire and Hertfordshire Regiment TA (55th Infantry Brigade, 18th British Division) arrives in Singapore and is rushed off to defend the north-east corner of the island, where it takes part in several skirmishes with the Japanese, losing thirty-two men.

FEBRUARY

Intensive fighting in Singapore from 8 to 15 February 1942 results in the capture of Singapore by the Japanese. About 80,000 British, Indian and Australian troops become POWs, joining 50,000 taken by the Japanese in the earlier Malayan campaign, and including 1,000 men from the 5th Battalion of the Bedfordshire and Hertfordshire Regiment and the Dunstable Artillery Territorial Unit, the 419th Battery of the 148th (the Bedfordshire Yeomanry) Field Regiment RA, also of the 18th British Division.

Churchill calls the ignominious fall of Singapore to the Japanese the 'worst disaster' and 'the largest capitulation in British military history'.

A special 'black propaganda' unit is set up at Milton Bryan near Woburn under Denis Sefton Delmer, who introduces a new concept of psychological warfare and conducts an impressive black propaganda offensive against the Third Reich.

APRIL

Queuing in a line not more than two abreast becomes compulsory if there are more than six people waiting for a bus.

JUNE

A 'hutment' is opened in Hockliffe Road, Leighton Buzzard, to house forty 'land girls' from the Women's Land Army. The first warden is Miss Whipp. It continues as a hostel until September 1950.

JULY

Convoy PQ17, the first Anglo-American naval operation in the war under British command, loses twenty-four of its thirty-five ships en route from Iceland to Arkhangelsk in Russia, when its escorting warships are ordered to turn back and the convoy to 'scatter'.

AUGUST

The Battle of Stalingrad: after a five-month siege the German 6th Army surrenders on 2 February 1943, in probably the most decisive action of the war.

NOVEMBER

An outbreak of mumps hits St Barnabas Infants School and by December it becomes an epidemic.

After a series of battles at El Alamein in Egypt, the Allied forces break the Axis line. Winston Churchill says of this victory: 'Now this is not the end. It is not even the beginning of the end, but it is, perhaps, the end of the beginning.' After the war, he writes: 'Before Alamein we never had a victory. After Alamein, we never had a defeat.'

1943

JANUARY

One of the last combined exercises scheduled for the Leighton Buzzard area, codenamed Vixen, is considerably curtailed due to bad weather. The icy conditions of the roads make it impossible for umpires and senior officials to arrive on time.

APRIL

First photo-interpretation of a rocket-propelled weapon (V1).

Another combined exercise, codenamed Lins, is held at Leighton Buzzard (20th) to test the military and civil defences. As in December 1941, the exercise shows a tendency for the various services involved to work separately.

JULY

Linslade police charge eleven cyclists in one week for riding without lights, all members of the armed services. In the following month, the court increases fines for the offence from 10s to £1 but this has little effect.

DECEMBER

First *Luftwaffe* use of rocket-firing fighters during Allied bombing raid on Berlin. These could shoot horizontally from several thousand yards away, according to a Lancaster gunner on the raid.

1944

JANUARY

Distribution of cod liver oil and orange juice begins for all children. Politicians see this as a pillar of the Welfare Food Programme.

JUNE

The D-Day landings at Normandy beaches begin on the 6th. After the invasion, 'Europe 60 Group', directed from Leighton Buzzard, lands in Normandy to set up a mobile chain of radar stations providing navigational cover for Allied aircraft operating in support of the ground forces as they advance through France and into Germany.

The first V1 'doodlebug' is launched, with London its target (12th), followed in September by the first V2, and streams of flying bombs are soon crossing the Channel. In all 3,000 V2s are aimed at London and other targets. Counter measures are introduced, including new detection equipment for No. 60 Group, and fewer flying bombs succeed in reaching London every day.

SEPTEMBER

Brussels liberated on the 2nd, followed two days later by Antwerp.

DECEMBER

In his 'Stand Down' speech, the king lauds the Home Guard as an example of patriotic honour. The men get no gratuities, but are allowed to keep their boots and battle dress.

1945

MARCH

One hundred and eighty-nine radar stations remain in operation around the coast of Britain, all controlled by 60 Group in Leighton Buzzard.

Field-Marshal Bernard Law Montgomery, commander-in-chief of 21 Army Group, issues instructions to all officers and men not to fraternise with Germans.

APRIL

Hitler commits suicide (30th).

There is a welcome return to illumination when Big Ben is lit up to signal the official end of the blackout (30th).

Black propaganda broadcasts from Milton Bryan are closed down.

MAY

VE Day is celebrated on the 8th. This and the following day are proclaimed national holidays. Schools, shops and businesses, except for those in the food trade, are closed.

AUGUST

VJ Day is celebrated on the 15th. Leighton Buzzard is again decked with flags and bunting for a two-day holiday.

OCTOBER

No. 60 Group at Oxendon House in Leighton Buzzard is disbanded and 'whisked out of existence with unseemly haste and little fanfare, given its indispensable contribution to victory'. In the victory souvenir number of *Radar Bulletin*, Air Marshal Sir James Milne Robb, the Commander-in-Chief of Fighter Command, states that nearly 5,000 enemy aircraft were destroyed as a direct result of the radar information supplied by No. 60 Group.

1939–45 WAR MEMORIAL KEY

		Location	War Memorial Code
Leighton Buzzard	War Memorial	Church Square	a
Leighton Buzzard	Beaudesert Book of Remembrance	Library, Lake Street	b
Leighton Buzzard	Cedars School Roll of Honour	Leighton Middle School	c
Leighton Buzzard	Baptist Chapel Brass Plaque	Hockliffe Street	d
Leighton Buzzard	Post Office	Church Square	e
Leighton Buzzard	Salvation Army	Lammas Walk	f
Leighton Buzzard	Cemetery	Vandyke Road	g
Linslade	War Memorial	Mentmore Road	h
Linslade	Roll of Honour	St Barnabas	i
Old Linslade	Churchyard and Cemetery	St Mary	j
....	....	....	
Bletchley			
(Fenny Stratford)	War Memorial	Queensway	k
Bletchley (Old)	War Memorial	Church Green Road	l
Bletchley (Old)	Roll of Honour	St Mary	M
Cheddington	War Memorial	St Giles	N
Dunstable	War Memorial	Priory church	O
Eaton Bray	War Memorial	St Mary the Virgin	P
Edlesborough (also Northall)	War Memorial	Leighton Road	Q
Eggington	War Memorial	St Michael	R

Heath and Reach	War Memorial	St Leonard	*S*
Luton	Vauxhall Motor Works War Memorial	Kimpton Road	*T*
Mentmore	War Memorial	St Mary the Virgin	*U*
Soulbury	War Memorial (from Wesleyan Chapel)	All Saints	*V*
Stanbridge	Roll of Honour	St John the Baptist	*W*
Stoke Hammond	War Memorial	St Luke	*X*
Wing	War Memorial	All Saints	*Y*
Leighton Buzzard Observer	Information from reports in newspaper		*Z*

Pitstone Memorial Hall and Tilsworth rolls of honour also exist: no local connections have been found with the names on the former; no wartime deaths found among those on the latter.

APPENDIX C

LOCAL MEMORIALS

Surname	Forenames	War Memorial Code
Abbott	Stanley John	*c-i-l*
Airey	William George	*V*
Alves	Thomas	*U*
Badcock	Frederick George	*a-b*
Bailey	Kenneth	*h-i*
Baines	Henry Richard	*a-b*
Bastin	Rudolphe Gaston	*a-b 1st Leighton Buzzard Scouts*
Billingham	Herbert George	*A*
Blew	A.V. (Albert)	*Y*
Bond	John	*S*
Boyce	Stanley	*Z*
Boyd-Thomson	Richard Glencairn	*v Canadian Virtual War Memorial*
Bray	Stanley	*a-f*
Brown	John	*Z*
Buchanan	Arthur Geoffrey	*Y*
Butler	Richard William	*a-c Lidlington*
Cato	Joseph	*v-x*
Chambers	George Harry	*T*
Chorlton	Geoffrey	*Q*
Clark	Albert James	*a-b-d*
Clarke	Frank Thomas	*no memorial, buried Heath & Reach*
Clayton	Herbert Thomas	*A*
Clifford	Seymour Fitzclare	*L*
Compton	Frederick William	*h-t*

Cook	Edward John	*P*
Cooper	Alfred John	*a-b Waterlow & Sons Ltd (Dunstable Cemetery)*
Cox	Reginald Thomas	*b-r*
Creighton	Albert Victor	*a-e*
Cunningham	Peter Chris	*c-q*
Davis	Percy James	*a-b*
Dawson	Samuel Henry	*N*
Day	Reginald Victor	*P*
Dell	William Joseph	*a-b*
Dennison	Ronald Henry	*a-b-o*
Deverell	John Edward Steadman	*c Tring*
Don	Alan McQueen	*Z*
Dorsett	George	*H*
Draper	George Oliver	*-*
Duckett	Cyril William	*Y*
Dyer	William Charles	*H*
Edmunds	Richard Arthur	*U*
Eggleton	Charles Dennis George	*a-b*
Ephgrave	Bertie	*L*
Fairfax	Thomas	*Z*
Faulkner	Kenneth Douglas	*c-h-i*
Fitsell	Richard Louis	*a-b-c*
Flemming	Sidney Cyril	*a-b*
Foster	Frederick John	*a-b-c*
Gadd	George Edward	*c-o*
Gale	Robert Frank	*X*
Garner	William George	*a-b*
Gaskin	Ernest Robert	*a-b-f*
Gibbins	George Thomas	*h-i*
Gilbert	Kenneth Harold	*a- b*
Gill	Ernest Hector	*A*
Goodway	Stanley Robert	*Y*
Gray	Richard Ernest	*p-q Waterlow & Sons Ltd, Dunstable Cemetery*

Green	Cyril	*V*
Griffin	Arthur Charles	*a-b-c-d*
Griffin	Donald Frank	*d-h-t*
Grotrian	Charles Herbert Brent	*Z*
Hack	William James	*S*
Hancox	Reginald Sidney	*q Toddington*
Hankins	Glyn	*c-l-m*
Hardy	Aubrey Frederick Robert	*p Waterlow & Sons Ltd (Dunstable Cemetery)*
Hart	Stanley Frederick	*h-i*
Heady	John Joseph O'Brien	*c-h-i Westminster Bank, Woburn Sands*
Heald	Howard Harvey Noel	*y-z Aspley Guise; Bedford Modern School*
Hedges	Marshall Claude	*W*
Hedges	Bernard Frederick Russell	*b-h-i Waterlow & Sons Ltd (Dunstable Cemetery)*
Hewitt	George Edward	*A*
Hoddinott	John Henry	*C*
Horn	Derrick Leslie Walter	*c Ivinghoe*
Horne	Dennis Alfred	*c-h-i*
Horne	Donald Edwin	*c-o*
Hucklebridge	Frank Haworth	*Z*
Humphries	Paul Silwood	*H*
Hunt	Percy Harold	*Q*
Hunt	William George	*Q*
Huxley	Henry George	*L*
Hyde	Sydney Thomas John	*a-b*
Jardine	William Patterson	*a-t*
Jeffs	Frederick George	*c-h-i-l*
Jordan	Lionel John	*a-b*
Kempster	Ernest Arthur David	*B*
Kempster	James Edward Thomas	*Z*
King	Charles Henry	*T*
King	Francis John	*A*
King	William James	*a-b*
Lancaster	John William	*A*

Law	Harold John	*a-b-c-t*
Lee	William Henry	*a-b*
Linney	Frederick	*a-b*
Main	Caruth (Charles)	*H*
Marks	Percy Clifford	*a-b*
Maunders	Maurice	*P*
Maunders	Owen Percy	*P*
Maynard	Thomas Henry	*Z*
Mead	Donald	*p-q*
Merry	William Haslam	*Y*
Miller	Thomas William Lound	*Towyn*
Missenden	Norman Mayo	*h-i*
Mordecai	James Robert	*P*
Morris	Alan Henry	*a-b*
Morse	Charles John	*A*
Morton	John James	*Q*
Murgett	Arthur William	*B*
Norman	Norman Royston	–
Olney	Kenneth Walter	*a-b*
Page	Dennis	*c-y*
Paxton	Douglas Francis	*Y*
Pearson	Frank William	*Q*
Peasland	James Powell	*Z*
Perry	Harold Ernest	*c-k*
Pickering	Frederick Henry	–
Pidgley	Peter William	*c-y*
Pratt	Stanley Walter	*a-c*
Proctor	James Arthur Edwin	*h-i-l*
Prudent	Vivian Jack	*a-c Chalgrave*
Puddephatt	Frank David	*S*
Rayner	Kenneth Roland	*h-i*
Rollings	Kenneth George	*P*
Rowe	Geoffrey Aveline	*a-d*
Saunders	Alan Richard	*c-n*
Sayell	Kenneth J	*h-i*

Sear	James Theodore	*B*
Sells	Victor George	*b-w*
Shand-Kydd	John Victor William	*N*
Sharratt	Robert	*Q*
Sifleet	Ernest George	*A*
Sims	Henry Percy	*A*
Smith	Alfred Stewart	*Y*
Smith	Henry George	*H*
Smith	Kenneth Ernest Douglas	*a-b*
Smith	Leslie Frederick	*c-l-m 1st Bletchley Scouts*
Smy	Walter George Z.	*–*
Smythe	Reginald R.I.	*buried Ivinghoe*
Stanbridge	Kenneth William	*b-w*
Stannard	Leslie Frank	*a-b Tempsford*
Stevens	Cyril Ernest	*a-b-d*
Stone	Frederick John	*S*
Syratt	John Henry	*Y*

GALLANTRY AWARDS

BEM	British Empire Medal		DSC	Distinguished Service Cross
DCM	Distinguished Conduct Medal		DSO	Distinguished Service Order
DFC	Distinguished Flying Cross		MID	Mentioned in Despatches
DFM	Distinguished Flying Medal		MC	Military Cross
DSM	Distinguished Service Medal		MM	Military Medal

Award	Rank	Forenames	**Surname**	Announcement
MC	Major Royal Corps of Signals	Arthur William	**Barron**	LBO – 02/10/45
MC	Major RA	Sir Bryan Cosmo	**Bonsor**	28/06/45
DFC	Pilot Officer 159 squadron	Walter Ernest	**Boteler**	07/12/44
DSM	Petty Officer Ordnance Mechanic 4th Class Serving on HMS *Belfast* when the *Scharnhorst* was sunk	Gerald Frederick Turney	**Burgess**	07/03/44
DFC DSO	Wing Commander 156 Squadron	Allan John Laird	**Craig**	21/07/44 20/10/44
Croix de Guerre	Sergeant	Frederick George	**Dillamore**	LBO – 17/08/43
DSO	Captain RN	John Wentworth	**Farquhar**	14/11/44
DFC	Flight Lieutenant RAFVR	Kenneth Douglas	**Faulkner**	11/08/42
DFC	Flight Lieutenant 142 Squadron	Jack	**Fenwick-Webb**	28/09/43
MC bar	Major RAMC	Robert 'Bobby' Frank	**Gale**	07/09/43 27/01/44
DFC	Flight Sergeant 153 Squadron	John Joseph O'Brien	**Heady**	

MC	Lieutenant Queen's Own Cameron Highlanders	John Colin	**Hamp**	31/12/42
MM	Private RAMC	Frederick	**Hesk**	13/09/45
MM	Trooper	William George	**Hunt**	*London Gazette,* 27/09/40
MBE	Lieutenant RE	Michael Vivian	**Keelan**	18/10/43
DFC	Captain RA	Peter Emil	**Kroyer**	26/07/45
MM	Lance Corporal Green Howards	Charles Victor	**Linney**	04/11/43
MID	Lieutenant-Commander RNVR	John Ivester	**Lloyd**	07/11/44
DSM	Leading Seaman LCH 185	Caruth (Charles)	**Main**	07/12/43
DCM	2nd Lieutenant Essex Regiment	Thomas William Lound	**Miller**	03/09/18
DFM	Flight Sergeant 97 squadron	Robert John	**Oates**	31/12/42
DSO	Major Welch Regiment	Henry Thompson	**Pepper**	01/03/45
DFC	Flying Officer 604 Squadron	Donald Walter	**Ray**	23/01/45
MID	Commander (Captain) *Oronsay*	Richard	**Roberts**	
DSM	Stoker P/Officer submariner HMS *Unbroken*	Frederick Charles	**Sharp**	28/09/43
BEM	Military Division Flight Sergeant WAAF	Nancy Mary	**Shepherd**	01/01/44
MM	Sergeant RA	George Henry	**Smith**	11/10/45
DFC	Pilot Officer RAFVR 218 Squadron	Leslie Frederick	**Smith**	14/05/1943 06/08/1943
BEM	Military Division Company QMSM Essex Regiment	Ernest Stanley	**Ward**	13/12/45
MM	Sergeant Beds & Herts Regiment	Lionel Charles	**Wickson**	03/07/45
DFC DSO	Wing Commander 109 Squadron	Charles Victor Douglas	**Willis**	20/01/42 31/03/44
MM	Sapper Essex Regiment Royal Engineers	Eric Sydney	**Woodward**	19/08/43
MID	Sapper	E.F.	**Wright**	LBO – 24/07/45

BIBLIOGRAPHY

Antrobus, Stuart, *We wouldn't have missed it for the world: The Women's Land Army in Bedfordshire 1939–1950* (Dunstable: Book Castle Publishing, Dunstable, 2008).

Argent, Private Denis, *A Soldier in Bedfordshire 1941–42: The diary of Private Denis Argent, Royal Engineers* (Woodridge: Bedfordshire Historical Record Society/Boydell Press, 2009). He kept a diary as part of the Mass Observation project started in 1937.

Audric, Brian and Omer Roucoux, 'The Meteorological Office Dunstable and the IDA Unit in World War II', occasional publication No. 2, the Royal Meteorological Society, Reading, 2000, pages 1–17.

Barrow, Lieutenant-Colonel T.J., Major J.A. French and J. Seabrook, *The Story of the Bedfordshire and Hertfordshire Regiment (The 16th Regiment of Foot) Volume II 1914–1958* (Farnham: History Committee, The Royal Anglian Regiment (Bedfordshire and Hertfordshire), 1986).

Beckett, Constance Mary, *The Sky Sweepers* (London: Regal Life Ltd, 1995).

Bennett, Arnold, *Tessa of Watling Street: A Fantasia on Modern Times* (London: Chatto and Windus, 1904). A detective novel comparing the slow pace of 'rustic' life in Hockliffe with the whirl of the London financial world. It mentions Leighton Buzzard train station several times. Bennett lived in Hockliffe 1901–03. One passage describes how 'the Chiltern Hills stretched away in the far distance, bathed in limitless glad sunshine; and Watling Street ran white, dazzling and serene, down the near slope and up the hill towards Dunstable, curtained in the dust of rural traffic'.

Booth, Dave, *The History of Linslade Wood* (Leighton Buzzard: Friends of Linslade Wood, 2011).

Bowman, Martin W., *The Bedford Triangle: US Undercover Operations from England in the Second World War* (Stroud: The History Press, 1996, 2003, 2009).

Brooks, R.J., *Thames Valley Airfields in the Second World War* (Newbury: Countryside Books, 2000) (information on Wing airfield).

Brown, Paul, Leighton Buzzard and Linslade, A History (Phillimore & Co., 2008). The history of this important medieval town right up to modern times. The town is sometimes, because of its name, used as the butt of jokes but, as this book shows, it has an important place in history.

Brown, Paul, 'Leighton Buzzard at War – personal recollections – Ludovic McRae and Viv Willis', *Transactions of Leighton Buzzard and District Archaeological and Historical Society*, no, 4, pp.13–14.

Bunker, Stephen, *The Spy Capital of Britain: Bedfordshire's Secret War 1939–1945* (Bedford: Bedford Chronicles, 2007).

Calder, Guinvere E., *The History of Eggington* (published by the author, 1986).

Carroll, David, *Dad's Army: The Home Guard 1940–1944* (Stroud: Sutton Publishing, 2002).

Carty, Pat, *Secret Squadrons of the Eighth* (London: Ian Allan Ltd, 1990).

Clarke, Freddie, *Agents By Moonlight: The Secret History of RAF Tempsford during World War II* (Stroud: Tempus, 1999).

Cutler, Jane, *Voices from an English Village: Cheddington – Memories of the Second World War*, compiled with assistance from Jean Fulton, John T. Smith and Denise Webb (Cheddington: Cheddington Historical Society, 2011).

Delmer, Denis Sefton, *Black Boomerang* (London: Secker & Warburg, 1962).

Delve, Ken, *The Military Airfields of Britain: The Northern Home Counties* (Marlborough: The Crowood Press Ltd, 2007).

Garnett, David, *The Secret History of the Political Warfare Executive 1939–1945* (London: St Ermin's Press, 2002).

Hart, Richard, *Pictures Past and Present* (Luton: White Crescent Press Ltd, 1986). A photographic history of Leighton Buzzard, Linslade and the surrounding district.

Howe, Ellic, *The Black Game* (London: Michael Joseph Ltd, 1982). Contains a chronological list of clandestine broadcasting stations operated by the Political Warfare Executive at Simpson and Milton Bryan.

Latham, Colin and Anne Stobbs, *Pioneers of Radar* (Stroud: Sutton Publishing, 1995).

Medley, R.H., *Cap Badge: The Story of Four Battalions of the Bedfordshire and Hertfordshire Regiment and the Hertfordshire Regiment (TA) 1939–1947* (London: Leo Cooper, 1995).

Nowodworski, Stanley Newcourt, *Black Propaganda in the Second World War* (Stroud: Sutton Publishing, 2005).

Olsen, Reed Oluf, *Two Eggs on my Plate* (London: The Companion Book Club, 1954). Originally published by George Allen & Unwin Ltd.

Parker, Ray, *Leighton Buzzard's Communications War* (unpublished manuscript, 2011).

Perquin, Jean-Louis, *The Clandestine Radio Operators* (Paris: Histoire & Collections, 2011).

Pether, John and Peter White, 'Political Warfare in Bedfordshire', *Bedfordshire Magazine*, vol. 26, no. 208, Spring 1999, pages 320–325.

Pigeon, Geoffrey, *The Secret Wireless War – the Story of MI6 Communications 1939–1945* (St Leonards-on-Sea: UPSO, 2003).

Schneider, Joan (compiler and editor), *'I can remember … village life recollected in the words of people from Tilsworth, Bedfordshire'* (Tilsworth: Friends of Tilsworth Church, 2006).

Radar Bulletin of RAF Group 60 HQ, 'Victory Souvenir Number', 1945.

Rankin, Nicholas, *A Genius for Deception: How Cunning Helped the British Win Two World Wars* (Oxford: Oxford University Press, 2008).

Rees, Nigel, *The Czech Connection – the Czechoslovak Government in Exile in London and Buckinghamshire* (published by the author, 2005).

Risby, Stephen, *Prisoners of War in Bedfordshire* (Stroud: Amberley Publishing, 2011).

Robinson, J.R., '60 Signals Group, Fighter Command, Royal Air Force', (Chapter XII of *The Canadians on Radar 1940–1945* website, www.rquirk.com).

Smith, Graham, *Hertfordshire and Bedfordshire Airfields of the Second World War* (Newbury: Countryside Books, 1999).

Smith, Michael, *Station X: The Codebreakers of Bletchley Park* (London: Channel 4 Books, 1998).

Spark, Muriel, *The Hothouse by the East River* (London: Macmillan, 1973). 'The Compound' in this novel was based on the Milton Bryan studio. Her most famous work, published in 1961, was *The Prime of Miss Jean Brodie*.

Taylor, John A., *Bletchley Park's Secret Sisters: Psychological Warfare in World War II* (Dunstable: The Book Castle, 2005).

Taylor, John A., *Bletchley and District at War: People and Places* (Copt Hewick: Book Castle Publishing, 2009).

Tipping, Peter (ed.), *Somewhere in England - the BBC in Bedford during World War II* (Bedford: Bedford Society, 2001).

Towler, Ray and Barry Abraham, 'Satellite Landing Grounds, Part 3', in *Airfield Review*, no. 40, April 2002, p.29. An account of the use of the airstrip at Woburn Park.

'W, Peter'. 'Memories of Political Warfare', Imperial War Museum, Misc 91, item 1334.

Warth, Michael, *Wings over Wing: The Story of a World War II Bomber Training Unit* (Dunstable: The Book Castle, 2001).

Webster, Leslie, 'Life in a G-String' (Leighton Buzzard: privately published, 2000, copyright Helen A. Webster). An account of a talk given by Sergeant Webster to Leighton Buzzard Rotary Club in 1988 about his experiences as a POW of the Japanese.

Willis, R.V., *Born to Build* (privately published).

Willis, R.V., *The Coming of a Town, the Story of Leighton Buzzard and Linslade* (Luton: White Crescent Press, 1984).

Wyatt, John with Cecil Lowry. *No Mercy from the Japanese* (Barnsley: Pen & Sword Books Ltd, 2008).

WW2 People's War. This is a fascinating online archive of wartime memories contributed by members of the public and gathered by the BBC. The archive can be found at bbc.co.uk/ww2peopleswar.

Yates, Jean and Sue King, *Dunstable and District at War* (Dunstable: The Book Castle, 2006).

Younghusband, Eileen, *One Woman's War* (Cardiff: Candy Jar Books Ltd, 2011). The author was awarded the British Empire Medal in the 2013 New Year Honours List for services to lifelong learning.

Footnotes in the History of Leighton Buzzard

In 1952 author Kathleen Mary Norton's most famous work *The Borrowers* was published. The story is about the adventures of tiny people who live underneath the floorboards. The location for the story, Firbank Hall, was based on Cedars House, Church Square, in Leighton Buzzard, where she had lived for a while as a child.

Captain Potts says to Sergeant Grimshawe in the 1958 film *Carry on Sergeant* (the very first 'Carry On' production), 'If you're ever Leighton Buzzard way, you know … feel like looking in … always pleased to see you.'

Leighton Buzzard railway station, actually in Linslade, was the location for part of the heavily fictionalised 1967 film *Robbery*, based on the so-called 'great train robbery' of 1963. The actual robbery took place just outside the town, at Bridego Bridge, Ledburn.

In the TV Programme *Room 101* in 2001, comedy duo Mel and Sue (Melanie Clare Sophie Giedroyc and Susan Elizabeth Perkins) chose Leighton Buzzard as one of their pet hates, because of a particularly disastrous gig they once did there. This caused controversy, reported in the *Leighton Buzzard Observer*, with the town council claiming that the programme did not have permission to use an image of the town's coat of arms.

INDEX

Bold numbers denote reference to illustrations.